# The Saturn V F-1 Engine

## Powering Apollo into History

Anthony Young

# The Saturn V F-1 Engine

## Powering Apollo into History

Published in association with
Praxis Publishing
Chichester, UK

Anthony Young
Space Historian
Miami
Florida
USA

SPRINGER–PRAXIS BOOKS IN SPACE EXPLORATION
SUBJECT *ADVISORY EDITOR*: John Mason B.Sc., M.Sc., Ph.D.

ISBN 978-0-387-09629-2 Springer Berlin Heidelberg New York

Springer is part of Springer-Science + Business Media (springer.com)

Library of Congress Control Number: 2008930113


Printed in Germany



Cover design: Jim Wilkie
Editor: David M. Harland
Typesetting: Originator Publishing Services, Gt Yarmouth, Norfolk, UK

# Contents

# Illustrations

## Color section

# Foreword

Power to boost the mighty Apollo and Skylab Saturn V launch vehicles required designing and developing a liquid propellant rocket engine with nearly four times the thrust of any rocket engine then in service in the free world. This new high thrust engine, designated the F-1, began design in 1958 and subsequently boosted 13 Saturn V launch vehicles with 100 percent reliability through 1973.

Anthony Young has captured the story of the F-1 engine design and development, through extensive documentation research and interviews with people still available, to provide the history of the F-1. Anthony not only presents the story of the F-1, but also much insight into early NASA and contractors' moon-landing launch vehicle trade studies. His story also includes a description of the role of NASA F-1 Project Management, Boeing S-IC stage development, and Saturn V launch vehicle stacking and processing. Of course, no F-1 story would be complete without coverage of the next generation F-1, the F-1A, which was in test with its 1.8 million lbs of sea level thrust before the Saturn V program was cancelled.

The F-1 story had its roots in the 1940's and 1950's as America developed missile and space launch vehicle systems. Power increased as well until the Rocketdyne E-1 engine system reached 400,000 lbs of sea level thrust. In the late 1950's, before NASA was formed, the Air Force and Rocketdyne engaged in studies to answer the question "What would be the maximum thrust of a rocket engine that would ever be required?" The study answer identified a thrust level of 1,000,000 lbs of sea level thrust. With this target thrust level, the Air Force gave Rocketdyne the go ahead and provided modest funding for design, production and testing of a 1,000,000 lb thrust solid wall thrust chamber. In keeping with the alphanumeric designation system it had established, Rocketdyne identified the 1,000,000 lb thrust engine as the F-1.

Concurrent with the time of the Air Force contract, the U.S. Congress established a civilian space agency and, on October 1, 1958, NASA came into being. Almost immediately the F-1 rocket engine tasks were transferred from the Air Force to NASA in keeping with the agency's charter. The target thrust had increased to 1.5 million lbs along the way. Development continued on the solid wall chamber, and, as the thrust chamber testing began, high noise levels rumbled across the mountains at the Santa Susana Field Laboratory such that the name 'King Kong' was used to describe the system.

During design and development of the F-1, the many challenges were met by a very determined and skilled workforce. Combustion instability, turbopump LOX pump failures, and minor problems like small cracks, erosion and leaks, had to be overcome. Just the very size of the F-1 engine required a lot of technical skills and equipment to handle, transport, and protect it during production, test and launch. The furnace brazing of the thrust chamber coolant tubes, nozzle extension (skirt) design, handling, and installation, and the thermal insulation system to protect the engine during launch and flight, all required many engineering innovations.

F-1 production was at Rocketdyne's Canoga Park facility, and engine level testing was primarily conducted at the Edwards Field Laboratory. Initially, three test stands were constructed there, but as testing increased three more stands were announced by NASA in 1962.

Rocketdyne assigned and relocated me with my family to MSFC as a participant in the early F-1 engine integration into the S-IC stage and its subsequent testing. Many exciting achievements were accomplished there over the next three years by the NASA, Boeing and Rocketdyne team. There were a lot of modification upgrades (approximately fifty) that had been identified during development at Rocketdyne. These could not be installed during the production process before engine delivery, and had to be installed in the initial test and flight engines after delivery to MSFC. Modifications such as welding studs onto the thrust chambers for attachment of the thermal insulation, and welding gussets on the thrust chamber for strengthening the support of the 'thrust-OK' pressure switches, were required. These were installed by Rocketdyne Traveling Production Teams. Other modifications like transducer and harness changes were accomplished by NASA technicians with Rocketdyne Field Engineering technical support, or by the single Rocketdyne technician assigned to MSFC. Rocktedyne also operated a large modification kit and support hardware warehouse at MSFC for NASA that we were all proud of, and that contributed to the processing success of the F-1 engines at MSFC.

On the S-IC-T test vehicle, the F-1 engines were installed vertically into the stage with the stage installed in the test stand. The first three tests of the S-IC-T were of the center engine only. The first two tests were premature cutoffs, with the first inadvertently cutoff by an observer and the second due to a broken wire in a safety circuit. The third test successful ran for approximately the programmed 15 seconds. The next test, conducted on April 16, 1965, was with all five engines running for the programmed duration of approximately seven seconds. The five F-1 engine cluster with a thrust of over 7.5 million lbs shook all of north Alabama when fired. What a thrill to watch and feel. I don't recall the actual distance, but the control center at MSFC in the Test Division seemed quite close to the test stand, and one got a good feel for the power during an F-1 engine cluster test. Communications then were not as electronically networked as they are today, and I had the thrill of relaying by telephone the countdown and the initial test firings from the control center to a room full of F-1 executives and engineers at Rocketdyne, Canoga Park.

The overall success of the F-1 program resulted from a strong team effort by NASA and Rocketdyne Program/Project Management, Engineering, Quality, Procurement, Production, Test and Launch Personnel. The F-1 engines remaining after completion of the Apollo and Skylab programs can be found today mounted on S-ICs at the U.S. Space & Rocket Center, Alabama, the Kennedy Space Center, Florida, and the Michoud Assembly Facility, Louisiana, and as standalone displays in several museums throughout the U.S. and some foreign countries. An F-1 engine stands proudly in front of the Rocketdyne (now Pratt & Whitney Rocketdyne) plant in Canoga Park, California. The engine was placed there on the tenth anniversary of the Apollo 11 lunar landing as a tribute to the men and women who designed and developed it.

*Vincent J. Wheelock*
Rocketdyne Director (retired)
Field Engineering & Logistics

# Author's preface

One of the most powerful machines ever conceived and built was Rocketdyne's F-1 engine. It had a dubious start—proposed for a rocket that did not exist, for a mission that was not defined. Nevertheless, the company began studies on a 1 million pound thrust rocket engine for the Air Force in the 1950s. When it took over the contract, NASA increased the thrust requirement to 1.5 million pounds. This was fortuitous, as it was instrumental in enabling America to achieve President Kennedy's challenge of landing astronauts on the Moon and returning them safely to Earth within the decade of the 1960s.

After completing *Lunar and Planetary Rovers: The Wheels of Apollo and the Quest for Mars* for Springer–Praxis, I pondered which subject relating to the Apollo program I could write about next. No book had been written on the Rocketdyne F-1 engine that powered the first stage of the Saturn V launch vehicle. I knew the story of this engine's development would involve human interest as well as engineering. I was mystified as to why no one had yet written a history of this superb engine with a 100 percent success record in flight. I decided to take up this challenge. Sadly, the passage of decades since the end of the F-1 engine program has taken its toll on the roster of men who worked on it at Rocketdyne and the Marshall Space Flight Center. Two key individuals whom I would have particularly liked to interview were David Aldrich, Rocketdyne F-1 Engine Program Manager, and his deputy, Dominic Sanchini, but both had passed away some years ago. I had hoped to interview Leland Belew, who managed the engine programs at MSFC, but owing to his age he was unavailable. There were others, both at Rocketdyne and NASA, whom I could not interview. As a result, this book is an imperfect history of the F-1 engine. However, there were sufficient engineers and managers both at Rocketdyne and NASA able to speak, and they provided information and materials that helped to fill the gaps.

While establishing the content for this book, I realized that the F-1 engine and the S-IC stage it powered were an integral system. I therefore chose to include chapters about the S-IC and to discuss the overall Apollo program management.

This actually helped to properly round out the F-1 engine program. If I had not done so, I am sure that I would have received queries from readers asking why I had not covered these subjects.

With the advent of solid rocket motor technology for the Titan rocket and Space Shuttle, and their forthcoming use by the Ares launch vehicles of the Constellation program, it is unlikely that a liquid propellant rocket engine of the size and power of the Rocketdyne F-1 will again see service—it will remain one of the greatest engineering achievements of the 20th century.

*Anthony Young*
May 2008

# Acknowledgements

Early in this project, I was fortunate to get the generous help of Rocketdyne veteran Vince Wheelock. He offered suggestions on chapter structure, volunteered to ferret out information from Rocketdyne's archives, gave me priceless documentation on the F-1 engine program (including rare articles, brochures, press releases and other information) and, above all, many of the photographs included in this book. His wife Gail assisted him in scanning the photos and putting them on CD for me to use. Mr. Wheelock also reviewed the chapters specific to the F-1 engine. He put me in touch with Rocketdyne engineers, some retired and others still active, most notably Ted Benham and Bob Biggs, as well as those with whom he worked while at the Marshall Space Flight Center. I am indebted to Mr. Wheelock for his tireless help in recording the history of the F-1 engine.

I found other Rocketdyne engineers by following threads on the internet. One of those was Dan Brevik, who shared with me at length some of the inner workings at Rocketdyne during the F-1 program. Another was Ernie Barrett, who worked to set up the test stands at the Propulsion Field Laboratory, later called the Santa Susana Field Laboratory.

My visit to Huntsville, Alabama in April 2007 was very fruitful. Archivist Anne Coleman at the M. Louis Salmon Library of the University of Alabama in Huntsville helped immensely by directing me to MSFC documentation on the F-1 program and Boeing S-IC stage. Mike Wright at the MSFC History Office helped me to retrieve some very rare F-1 engine documentation and photographs that I could find nowhere else. My interviews in Huntsville were also very productive. In particular Saverio "Sonny" Morea, F-1 Engine Project Manager at MSFC; his deputy, Frank Stewart; and Richard Brown and Ron Bledsoe, who worked in the Engine Program Office; Konrad Dannenberg of the Technical Liaison Office; and Bill Sneed, who worked in the Saturn V Program Office.

Heinz-Hermann Koelle was an instrumental member of Dr. Wernher von Braun's team in Huntsville from the late 1950s until 1965 as Chief of Preliminary

Design for the Army Ballistic Missile Agency, and subsequently Director of the Future Projects Office at MSFC reporting directly to Dr. von Braun. Prof. Koelle reviewed the first two chapters of this book, and his comments were most helpful.

I had the very good fortune to learn about Harold C. Hall, who worked in the Engine Program Office during the 1960s, during which time he compiled binders of photographs and photo descriptions routinely sent to the F-1 Engine Program Office at MSFC to record progress with the engine's development. Hall had organized this material by year for the entire life of the program at MSFC. These binders had been in storage for 40 years until I visited Mr. Hall in Huntsville. He generously loaned them to me in order that I could scan some of the photos for this book.

Shelly Kelly, the University Archivist at the Neumann Library of the University of Houston at Clear Lake in Houston, Texas, was of immense help. Ms. Kelly pulled all the relevant taped interview transcripts conducted by Roger Bilstein for his book, *Stages to Saturn*. Bilstein rarely used quoted passages from these interviews. I found a wealth of information in the transcripts relating to the F-1 engine and S-IC stage, and have made use of it with appropriate annotation.

Fellow author Dr. David M. Harland edited the manuscript for Praxis, for which I am most grateful.

# Introduction

As the train made its way southwest across Texas, Dr. Wernher von Braun occasionally looked out the window in curiosity. The flatlands were totally unlike the landscape of Germany and Bavaria that he had left behind some months earlier. The saga of how he led many of the engineers, technicians and scientists of his team from Peenemünde to surrender to the U.S. Army's 44th Division in the final days of the war in Europe as part of Operation Overcast would be recalled in future books and articles. For now, however, he and the others were part of a top secret program, now codenamed Operation Paperclip, to tap their collective technical genius and apply it to American rocketry.

With von Braun was Maj. James P. Hamill of the Office Chief of Ordnance, Sub-Office, Rocket, and Capt. William E. Winterstein, Commanding Officer of the 9330 Ordnance Technical Service Unit, Sub-Office, Rocket. Hamill had been charged by Col. Holger Toftoy, Chief of Army Ordnance Technical Intelligence, to transfer the first wave of these Germans safely from Europe to Ft. Bliss, Texas, along with V-2 components, machine tools, instruments and documentation. Von Braun's team were a scientific and engineering treasure trove, and Toftoy wished to ensure their identity, the work they had performed on Germany's rocket programs, and their new duties in the U.S., remained absolutely secret. It was Hamill's job to prevent information 'leaks' and to make sure that none of the Germans went missing in transit. Because anti-German sentiment in America was understandably high as a result of the war, von Braun passed himself off as being Swiss, and his colleagues took on a variety of nationalities and occupational aliases. Their families had been temporarily housed in Landshut, Germany, and if all went well they would be shipped over toward the end of 1946. At Ft. Bliss the underutilized William Beaumont Army Hospital Annex was to be converted into living quarters for the families. In fact, the Operation Paperclip Germans were under contract to the U.S. Government for a period of six months. If after this time a contract was not extended, the man would be returned to Germany. Von Braun doubted that he would be sent back to Germany, however.

At the close of World War II, Operation Paperclip secretly brought many Germans who had worked on the V-2 to America to work on American missile and rocket development, and eventually the Saturn V. Front row, fourth from the left is Dr. Arthur Rudolph. Front row, seventh from the right is Dr. Wernher von Braun. (White Sands Missile Range)

His country had been ravaged by war, many of its cities were in ruins, and its infrastructure was decimated; to rebuild it would take years, with all the shortages that this would entail. He felt his prospects were better in America, and he looked forward to having his sweetheart Maria von Quistorp join him in order that they could be married.

Von Braun was a rocket engineer first and foremost. At Peenemünde on the north coast of Germany his team had worked to design, test and build the A-4/V-2 rocket. Although it was made clear that his task now was to advance American missiles, he explained to anyone who would listen that it would be possible to use rockets to explore 'outer space'. When the train halted in El Paso that October day in 1945, he disembarked and stood with Maj. Hamill, and the others congregated to await their transport to Ft. Bliss. In order not to draw undue attention, the team of 120 men was traveling in small groups, and this was the first. It included Dr. Walter Dornberger, Dr. Kurt Debus and Dr. Eberhard Rees, who would all go on to profoundly influence America's future in space exploration. An American concern was that an even larger number of Germans, including specialists in jet propulsion and nuclear physics, had ended up in Russia. This divided German brains' trust would provide the impetus for the United States and the Soviet Union to compete in both the build up of missiles and the exploration of space.

The 9330 Ordnance Technical Unit was responsible for the security, housing and general welfare of the Germans at Ft. Bliss. The initial accommodations were rather spartan, but starting in the fall of 1945 the Germans began taking up residence in the remodeled Beaumont Hospital Annex buildings. The Army Ordnance Department had hired General Electric, aided by the Germans, to assemble, checkout and launch V-2s at the White Sands Proving Grounds in New Mexico, which had been activated on July 13, 1945. It is not true that completed V-2s were shipped from Germany. At the time of the capture of the production facilities at Peenemünde and Nordhausen, many rockets were in various states of assembly. At White Sands, therefore, many crucial components were either in short supply or were completely unavailable. For example, each rocket's guidance system needed two gyroscopes, but only fifty were brought to America. General Electric had no option but to reverse-engineer many of the components, including gyros, servo motors, electrical distribution panels, wiring, propellant piping etc. Every V-2 component from Germany had to be inspected, and either refurbished or replaced. In addition to working with the Americans to prepare and launch V-2s, von Braun began a research and development program to improve the performance of the rocket.

In fact, the U.S. had begun to fund the development of rockets in the mid-1930s, and had progressed far beyond the pioneering experiments of Robert H. Goddard. In 1936 Dr. Theodore von Karman at the California Institute of Technology (CalTech) started work at the Guggenheim Aeronautical Laboratory of the California Institute of Technology (GALCIT) to develop a rocket to 'sound' the upper atmosphere. In July 1939, he formally initiated the Rocket Research Project; the nation's first center devoted to propulsion systems. Their first project was to develop Jet-Assisted Take-Off (JATO) solid rocket motors for aircraft. The first demonstration was performed in July 1941. Liquid propellant JATO rocket motors followed. Von Karman's team

The WAC Corporal was America's first high-altitude 'sounding' rocket. It had a short duration solid propellant booster (visible on the left) and liquid propellant main stage. It was first launched at the White Sands Proving Grounds in New Mexico in 1945. (White Sands Missile Range)

created the Aerojet Engineering Corporation in 1942 to manufacture JATO for the Army. In 1943, the Army Ordnance Department urged CalTech to expand its work in rocket propulsion, and in 1944 Project ORDCIT (Project Ordnance—California Institute of Technology) was begun. In November 1944 the Rocket Research Project was reorganized and renamed the Jet Propulsion Laboratory (JPL). The first rocket

JPL developed was the Private. It was followed by the Corporal, which employed an Aerojet engine that burned a hypergolic mix of alcohol as fuel and fuming nitric acid with aniline as the oxidizer. Development then began on the smaller WAC Corporal. The fact that it was assisted at liftoff by a cluster of small solid rocket motors made it the first two-stage rocket made in the United States. It was to conduct atmospheric research. It had no attitude control system, but used fins for stability. The first fully operational WAC Corporal launch was at White Sands on October 11, 1945, and the rocket achieved an altitude of 70 kilometers.

An interesting development in the fall of 1945 was the assistance and support of the U.S. Navy Bureau of Ordnance at White Sands. The Naval Ordnance Missile Test Center was established there in 1946. The synergy of the two branches of the military with CalTech and von Braun's team accelerated the development of rockets in America. Space is defined as starting at 80 kilometers, and on March 22, 1946 a WAC Corporal attained that altitude. This accomplishment was made against a backdrop of research into high energy rocket fuels, both by JPL and others. In 1945, the Navy's Bureau of Aeronautics had set up the Committee for Evaluating the Feasibility of Space Rocketry. It focused on single-stage-to-orbit research, with hydrogen and oxygen being viewed as the most likely combination of propellants. In their report entitled *Investigation on the Possibility of Establishing a Space Ship in Orbit Above the Earth's Surface* in November 1945, Lt. Cmdr. Otis E. Lancaster and J.R. Moore concluded that although single-stage-to-orbit was impracticable, a multi-stage rocket powered by hydrogen and oxygen could put a satellite into orbit. (History would prove the validity of multi-staging.) JPL evaluated this report, and in July 1946 recommended that Aerojet conduct research and development of such an engine. These pioneering investigations would ultimately lead to the development of the upper stages of the Saturn V.

The U.S. Army Air Force was also interested in exploring rocket research. Donald Douglas, President of the Douglas Aircraft Company, proposed that the Air Force establish a research and development organization for advanced propulsion in aircraft and rockets. The result was the Project RAND brains' trust, created in 1946. On the basis of the largest rocket so far developed, the V-2, it assessed the potential for a variety of propellants and studied the possibility of a satellite. Its report entitled *Preliminary Design of an Experimental World-Circling Space Ship* concluded that a multi-stage rocket powered by liquid oxygen with either hydrogen or alcohol could orbit a payload.

Both the Navy and the Army Air Force (the Air Force would separate from the Army in 1947) were motivated to pursue the rocket research recommended by these reports, but in the immediate post-war environment Congress was unwilling to make available the large amount of money that would be required. Instead, more modest programs involving less exotic propellants would be the order of the day for several years to come. Central to these development missiles and upper atmospheric studies were the Private A, Private F, Corporal, WAC Corporal and the V-2s modified at White Sands—as well as the first two-stage liquid propellant rocket in which a WAC was mounted on top of a V-2 in a configuration known as Bumper WAC. The first V-2 firing at White Sands was a captive, 57-second live-fire test on March 15, 1946.

The V-2 rockets launched from the White Sands Proving Grounds in the late 1940s and early 1950s were built from refurbished and American-made components. (White Sands Missile Range)

Gen. John B. Medaris was the first Commanding General of the U.S. Army Ballistic Missile Agency at the Redstone Arsenal, Alabama. (Redstone Arsenal Historical Information)

Four weeks later, the first V-2 flight abruptly ended after climbing to an altitude of only 5.5 kilometers. However, on May 10 the third rocket achieved an altitude of more than 112 kilometers, with instruments from the Applied Physics Laboratory of Johns Hopkins University to measure cosmic radiation. Thereafter, launches were made two to three times per month. By December 1946, the 17th V-2 reached an altitude of over 186 kilometers.

Another pivotal event took place in 1946. Capt. Winterstein had been in contact with von Braun, and many of his team, almost on a daily basis that year. He would occasionally invite von Braun and several members of the team to his home, and his wife would serve a home-cooked meal that was much appreciated. The conversation often centered on space travel, and the possibility of flying to the Moon. During one of these dinners, Winterstein asked von Braun how much money and time would be involved in a program to put a man on the Moon and return him to Earth. Von Braun said that he would make some calculations. A few weeks later, he told Winterstein it would cost $3 billion and take a decade. This timetable would prove prophetically accurate, although the cost was rather optimistic. By this point, Winterstein knew von Braun well, and was aware that the brilliant Germans could serve America well. In particular, he felt that their contributions to America's fledgling rocket program would be vital to national security and technical progress. As Winterstein told this author in 2006:

> I was struck by von Braun's dynamic leadership—the way he talked and acted. He had the charisma of a true leader. I was soon convinced he was something that

> was good for America then, and good for America later. Here was this German rocket team still technically enemy aliens. They fully knew we were going to have problems with Russia, because they knew of the turn over of the V-2 factory [at Nordhausen] to the Russians from the Potsdam Agreement. They were very anxious to get started on missiles for defense, but the Congress said 'No.'
>
> I remember a conversation I had with von Braun one evening in 1946 at the La Hacienda Restaurant south of Las Cruces, New Mexico. He was going to get married in the spring, and he was looking for financial security. He asked me what the outlook was at Ft. Bliss for rocket research, and I told him it was bad news; which was the truth, I didn't lie to him. He said, 'Well, after we get out from under this Army surveillance I am thinking of going to private industry.' I had to agree he could make a lot more money that way, but I said, 'Wernher, if you go to private industry you can kiss your trip to the Moon goodbye. Things will probably turn better later on. If it ever comes around to the point where man goes into space, the Army will be the best place to be. You'll probably be known as the top rocket scientist in America and possibly the world. How about sticking with the Army?'
>
> The way I saw it, the rocket team had to stay together as a team. He listened to me and decided to keep the team together, even though every member of the team that stayed with the Army sacrificed a small fortune, knowing they could have made a lot more money going to private industry.

In 1947, with the combined efforts of Douglas Aircraft, JPL, General Electric and von Braun's team at Ft. Bliss and White Sands, research and development began on the Bumper WAC. In May 1948, Bumper No. 1 was successfully launched, with the dummy second stage achieving separation from its booster. On February 24, 1949 Bumper WAC No. 5 reached the impressive height of 393 kilometers and the WAC attained a maximum speed of 8,867 kilometers per hour.

The German rocket team had spent nearly four years in the New Mexico desert. It had become a metaphor for their technical skills and engineering creativity. But this was about to change. In 1948, the Army's Chief of Ordnance had designated the Redstone Arsenal in Huntsville, Alabama as a center for rocket and missile research and development. Construction was to start in the near future on research and testing facilities and housing. That same year, a public statement by Col. Toftoy had greatly encouraged von Braun. It reminded him of the prophetic advice by Capt. Winterstein two years before. "It is possible," Col. Toftoy stated, "this generation will see huge rocket ships carrying passengers, that can circle the Moon and return to Earth safely. If work could begin on such a project immediately, and enough money to finance it in the interests of pure science, it could be done and witnessed by persons who are alive today." On October 28, 1949, the Secretary of the Army approved the transfer of the Ordnance Research and Development Division, Sub-Office, Rocket, from Ft. Bliss to the Redstone Arsenal, with a change in the name to the Ordnance Guided Missile Center. The transfer would start in April the following year. White Sands Proving Grounds would remain just that, but the majority of the Army's rocket and missile research and development would now take place in Huntsville.

Dr. Wernher von Braun (center) in Huntsville with the directors of the newly formed laboratories at the Marshall Space Flight Center in 1960. From left, Dr. Ernst Stuhlinger, Dr. Helmut Hoelzer, Karl Heimburg, Dr. Ernst Geissler, Erich Neubert, Von Braun, William Mrazek, Hans Hueter, Eberhard Rees, Dr. Kurt Debus, and Hans Maus. (NASA/MSFC)

After the Soviet Union detonated its first atomic bomb in August 1949 and the war in Korea broke out in 1950, the pace of America's development of rockets and missiles accelerated. Although some members of von Braun's team did opt to take professions in private industry, the majority remained and their patience proved out. The early 1950s were marked by the development of new rockets and missiles for the Army that included the Redstone and Jupiter, and by the deployment of the Nike surface to air missile. Meanwhile, the Navy developed its Viking rocket and in 1951 the Air Force began Project MX-1593, which led to the Atlas intercontinental-range ballistic missile.

But the exploration of space was never far from von Braun's mind, and he seized every opportunity to promote its possibilities. In the 1940s and 1950s, the acclaimed space illustrator Chesley Bonestell inspired a generation with the stark beauty of the Moon, the planets and the prospect of space travel. In 1949 Bonestell and engineer-turned-writer Willy Ley collaborated on the book *The Conquest of Space*. It became a bestseller and, by inspiring thousands of young men and women to pursue careers in aerospace, had a major impact on America's future in space. Bonestell's paintings of orbiting space stations, traveling to the Moon and the surfaces of the planets were rendered in such exquisite detail, and the text, which was backed up by hard science that reflected emerging technology, was so believable, that the book instilled a sense of wonder and belief that the exploration of space would indeed be possible within the lifetime of its readers. At the invitation of Cornelius Ryan, editor of the popular *Collier's* magazine, in 1951 Bonestell attended a symposium in San Antonio, Texas on the topic of space flight. There, he met von Braun and other rocket scientists. The March 22, 1952 issue of *Collier's* included an article in which artwork by Bonestell illustrated exposition by von Braun. It was the first of a series. The October 18, 1952 issue featured the cover story *Man on the Moon: Scientists Tell How We Can Land There in Our Lifetime*. To follow up this collaboration, the book *Conquest of the Moon* was published in 1953. It described how bases would be established, and how explorers would travel about on the Moon's surface.

By 1953, von Braun had become Chief, Guided Missile Development Division, Ordnance Missile Laboratory at the Redstone Arsenal. In September 1954 he wrote a proposal called *A Minimum Satellite Vehicle Based Upon Components Available from Missile Development of the Army Ordnance Corps*. No nation on Earth had yet placed a satellite into Earth orbit, but von Braun knew that America could do so, and he recognized the significance of America being the first to achieve this task. In the proposal, he wrote:

> The establishment of a man-made satellite, no matter how humble, e.g. five pounds, would be a scientific achievement of tremendous impact. Since it is a project that could be realized within a few years with rocket and guided missile experience available *now*, it is only logical to assume that other countries could do the same. *It would be a blow to U.S. prestige if we did not do it first.*

His warning was prophetic. America's satellite program came under the auspices of the International Geophysical Year project of the National Academy of Sciences. In the mid-1950s, there was no sense of urgency in the U.S. to launch a satellite by using whichever rocket might be capable of achieving the task, and, as events would prove, this apathy would later cost the nation dearly.

Von Braun eagerly pursued other ways of getting his message across. One outlet was Walt Disney, who made a series of TV shows devoted to space exploration. The first, *Man In Space*, was broadcast on March 9, 1955. *Man and the Moon* aired later in the year. The final of these three memorable programs, *Mars and Beyond*, was in 1957. The estimated audience of about 40 million viewers made von Braun a household name in the U.S.

On February 1, 1956 the Army Ballistic Missile Agency was established at the Redstone Arsenal, with Gen. John B. Medaris in command. The Army Rocket and Guided Missile Center was established at the same time to handle other research and development programs at the Arsenal, with Gen. Toftoy in command. Intelligence was available that the Soviet Union was preparing to launch a satellite. This prospect had always concerned von Braun. He continued to lobby for permission to prepare a satellite, but the fledgling ABMA was rebuffed by Washington. The official U.S. effort to launch a satellite to mark the International Geophysical Year was being run by the Navy, using a multi-stage rocket named Vanguard. Unfortunately, this rocket was complex and its development was behind schedule. In January 1957 the Army's Chief of Research and Development, Lt. Gen. James M. Gavin, sought information from ABMA on the possibility of using the Jupiter-C missile as a satellite launch vehicle. In April ABMA advised Lt. Gen. Gavin that as a backup to the troubled Vanguard program a Jupiter-C could launch a satellite as soon as September 1957. The first launch of a Jupiter-C had been on September 20, 1956 and it had

Dr. von Braun photographed with (from left) Maj. Gen. Holger Toftoy, Dr. Ernst Stuhlinger, Hermann Oberth and Dr. Robert Lusser. (U.S. Army)

boosted a 14 kg payload representing a dummy satellite to a height of 1,100 kilometers while traveling in an arc some 5,300 kilometers across the Atlantic from Cape Canaveral in Florida. However, the Pentagon did not want ABMA in the satellite launching business, and Gen. A.P. O'Meara went to Huntsville to make it absolutely clear that a Jupiter-C was not to 'accidentally' place anything into orbit!

On October 4, 1957 the Soviets put Sputnik into orbit. The Cold War between the ideological enemies made itself felt directly in American's homes as TV newscasts described the stunning success of Sputnik. The *New York Times* headline read:

> *SOVIET FIRES EARTH SATELLITE INTO SPACE; IT IS CIRCLING THE GLOBE AT 18,000 M.P.H.; SPHERE TRACKED IN 4 CROSSINGS OVER U.S.*

On November 3 the Soviets launched a bigger and much heavier satellite carrying a dog named Laika, showing that in addition to having a powerful booster they also had the technology to keep an animal alive in the vacuum of space.

In response, the first Vanguard was rushed to completion. The attempt to launch it on December 6 was a total failure. Carrying a satellite weighing only 2 kilograms, the rocket rose less than a meter before losing power, falling back onto the pad and exploding. As a result of this humiliating defeat, the launch of a satellite became one of the highest national priorities. The reluctance of the Eisenhower administration to use an overtly military rocket to do so (and especially one built by the German team) evaporated. ABMA was directed to prepare a Jupiter-C and launch a satellite as soon as possible. Exploiting the outside interest in its expertise, on December 10 ABMA submitted a report to the Department of Defense entitled *Proposal for a National Integrated Missile and Space Vehicle Development Program*. This historic document suggested the need for a booster of 1.5 million pounds thrust. Meanwhile, preparations were underway to have Jupiter-C No. 29 launch a satellite built by JPL. It lifted off at 10:48 PM on January 31, 1958 and successfully placed the satellite, named *Explorer 1*, into orbit. Its Geiger counter revealed the presence of an intense belt of radiation high above Earth, now named after James Van Allen, the satellite's principal scientist. The space race was on! America was on the threshold of a bold, new era of exploration that would one day take its astronauts to the Moon.

# 1

# Evolution of the Moon rocket

When searching for the origins of what became the Saturn V launch vehicle for the Apollo program, powered by a cluster of five F-1 engines, one finds diverse threads. The evolution of the largest rocket in the 'free world' was not linear. Studies were performed in the late 1950s and early 1960s by the Army Ballistic Missile Agency (ABMA), the Advanced Research Projects Agency (ARPA), the Department of Defense, NASA-affiliated centers, U.S. Congressional committees, and contracted private aerospace firms. These exploratory efforts were both military and civilian. They involved both liquid and solid propellants. The size and power of the launch vehicle were a reflection of the different means of achieving a manned lunar landing and returning the crew safely to Earth. The primary methods of achieving this were a direct ascent from Earth and decent to the Moon; rendezvousing spacecraft above the Earth prior to departing for the Moon—referred to as Earth orbit rendezvous (EOR); or separating a two-man landing module from the main spacecraft in lunar orbit, landing on the Moon, returning several days later and rendezvousing with the main spacecraft for the return to Earth—known as lunar orbit rendezvous (LOR). All of these studies were essential prerequisites for hardware definition in advance of committing tens of billions of dollars. Once the method was selected, what would be the best launch vehicle configuration to achieve it? There were advocates as well as critics of each method and launch vehicle configuration. This discourse and debate was necessary for not only the rocket itself but also the long range direction of the infrastructure on Earth, and to assure the greatest prospect of mission success and crew safety.

The Soviet Union had demonstrated its heavy payload lifting capability in 1957, and was working to develop larger boosters. Months before the stunning news of Sputnik I and II, the Preliminary Design Section of Dr. Wernher von Braun's team at ABMA had initiated studies for a comprehensive rocket and missile development program for the United States. *Proposal: A National Integrated Missile and Space Vehicle Development Program* was published December 19, 1957, and presented to

1-1 Heinz-Hermann Koelle (right) was Chief of Preliminary Design at the Army Ballistic Missile Agency, and later Director of the Future Projects Office at the Marshall Space Flight Center, reporting directly to Dr. von Braun. Koelle was instrumental in helping to establish the configuration of the Saturn I, Saturn V and proposed Nova launch vehicles. (NASA, H.-H. Koelle Collection)

the Department of Defense. Heinz-Hermann Koelle, Chief of the Preliminary Design Section, would prove a key contributor to large launch vehicle configuration design. Within the proposal was the recommendation to develop a launch vehicle having 1.5 million pounds of thrust. This was to be done by clustering four E-1 engines burning RP-1 (refined kerosene) with liquid oxygen (LOX), each delivering almost 1.7 MN (379,837 pounds) of thrust. The Rocketdyne division of North American Aviation had begun development of the E-1 in 1955, and it had been successfully test-fired. In order to minimize the tooling design for the manufacture of the propellant tanks for the launch vehicle, the proposal called for using the propellant tanks of the Jupiter and Redstone missiles. The clustered tanks that made up the first stage could be test-fired on a test stand at the Redstone Arsenal. This booster was the largest of five rockets envisaged for the Juno family of launchers, although as events transpired the Juno III and Juno IV would remain only proposals. After the necessary propellant loadings and projected trajectories for launching satellites had been determined, and all the other necessary calculations made, a 1/10th scale drawing of this booster was made by Fritz Pauli in the Preliminary Design Section. However, as yet there were no funds to develop this rocket.

In the light of the national humiliation caused by the success of the Sputniks, this proposal prompted the Department of Defense to consider establishing a small, fast-moving research and development office capable of conducting programs to develop new technologies in various disciplines. ARPA was established by Department of Defense directive 5105.15, signed on February 7, 1958 by Secretary of Defense Neil H. McElroy. ARPA was made responsible "for the direction or performance of such advanced projects in the field of research and development as the Secretary of Defense shall, from time to time, designate by individual project or by category." Roy W. Johnson, a civilian and former vice-president of General Electric, was selected to direct the new military agency. This proved to be a very wise choice, and Johnson earned the respect of military personnel. Gen. Medaris in particular, praised his ability to run the fledgling agency. ARPA would play a significant role in the development of America's first large booster.

In January 1958, Gen. Medaris testified to Congress on the status of the nation's space activities. In addition to the missile programs under his responsibility, which included the new solid propellant Pershing, Medaris was also preoccupied with the need for a large booster. "As the weeks and months of 1958 flashed by," he wrote in his autobiography *Countdown for Decision*, "everyone was becoming increasingly concerned over the fact that there was nothing on our immediate horizon that looked like a space vehicle big enough to challenge the Russian lead or carry really significant loads. In January [1958] in my first appearance before Congress in connection with U.S. space activities, I had made the flat statement that the U.S. must command at least 1 million pounds of take-off thrust by 1961, or by 1962 at the latest, if we were to have any chance of overtaking the Russians."

The late 1950s saw many space exploration proposals, and another one appeared in February 1958 from the Lewis Aeronautical Laboratory of the National Advisory Committee for Aeronautics (NACA) in Ohio, led by Dr. Abe Silverstein. The chief goal of *A Program for Expansion of NACA Research in Space Flight Technology* was "to provide basic research in support of the development of manned satellites and the travel of man to the moon and nearby planets." Lewis would indeed become involved in such research, but it would be as part of NASA.

In April 1958 the Ballistic Missile Division of the U.S. Air Force published the first development plan for a manned military space systems program. The objective was to "achieve an early capability to land a man on the moon and return him safely to earth." This was perhaps the first time this particular phrase was used by the U.S. Government. The program was to be given the highest national priority, placing it on a par with the development of ballistic missiles. There were to be four phases, with robotic exploration leading to manned lunar landing missions by December 1965. It was clear that the Army and the Air Force would be vying for money to pursue their individual man-in-space plans. In 1958 the Air Force had hired Rocketdyne to do the preliminary design of a single-chamber rocket engine burning RP-1 and LOX able to develop 1 to 1.5 million pounds of thrust—which was as much as the Army hoped to achieve using a cluster of E-1 engines. Rocketdyne designated this new engine the F-1.

Dr. von Braun was also a member of the NACA special sub-committee on space technology, and in April 1958 he submitted *Interim Report to the National Advisory Committee for Aeronautics, Special Committee on Space Technology: A National Integrated Missile and Space Vehicle Development Program by the Working Group on a Vehicular Program*. Derived from the launch vehicle proposal submitted to the Department of Defense four months earlier, this laid out a space launch program spanning more than two decades, an array of launch vehicles, and grandiose plans that included interplanetary probes, flights to the Moon, missions to Mars and even Venus. It was detailed and considerably grander in scope than the document issued by one of the NACA's own laboratories two months previously. It also came with an estimated cost of $30 billion. Many of the NACA people who reviewed the proposal were aghast. Indeed, Dr. Hugh L. Dryden, head of the NACA, moved to distance his committee from it by simply forbidding its release by anyone at the NACA without the disclaimer that it was not officially endorsed by the NACA. However, this report did contain the recommendation for a large, single-chamber rocket engine capable of in excess of 1 million pounds of thrust. Even more interestingly, it recommended a manned lunar mission within the next 10 years. Three months later, in July 1958, a revised and less ambitious report was issued by the NACA Working Group on a Vehicular Program. Curiously, the recommendation to develop an engine capable of in excess of 1 million pounds thrust was eliminated. Instead, the document stated, "A development program [should] be initiated immediately for a booster in the 1.5 million pound (6.7 MN) thrust class, with emphasis on early availability."

At ARPA, Richard B. Canright, David Young and Richard S. Cesaro had discussed using existing rocket engines in a dense cluster for the first stage of a large launch vehicle. Clustering was advantageous for redundancy using proven engines of known reliability. Canright, an engineer at the Douglas Aircraft Company, went to Huntsville and met with Gen. Medaris and Dr. von Braun to discuss the use of up to eight engines in a cluster. Medaris expressed to Canright his initial doubts about getting so many rocket engines to perform together reliably. Von Braun and Medaris preferred the cluster of four E-1 engines. Canright said it was ARPA's view that the booster should be powered by modified Jupiter engines—either that or a different agency would be found to perform the work. A subsequent meeting took place at the Pentagon with Gen. Medaris and Dr. von Braun for ABMA and Roy Johnson, David Young and Richard Cesaro for ARPA. Von Braun was pragmatic, and felt that even although ABMA had never developed anything approaching its complex-

1-2 Dr. Wernher von Braun briefs President Dwight D. Eisenhower on the Saturn I at the Marshall Space Flight Center in September 1960. (NASA/MSFC)

ity, such a cluster of engines could probably be made to work. Medaris restated his reservations about so many engines working reliably, but was eventually won over. He explained in his autobiography at length the advantages of clustering engines:

All these engines, firing simultaneously, would enable us to put a busload of astronauts in orbit, achieve a soft landing on the moon, and accomplish many other desirable objectives, both peaceful and military.

Although there are many complicating factors, there are also a host of good reasons for clustering several smaller motors to get a lot of power.

First of all, clustering engines makes it possible to take advantage of the considerations of safety and reliability that have led to the universal adoption of the multi-engine airplane as the safest mode of air transport. Thus, if the capabilities of the vehicle are not stretched to the limit, and something less than the last pound of possible weight is carried, a space mission can be completed with one or more engines out of action or functioning at less than full power.

Next, there is the question of control. Driving a missile up through the atmosphere is like balancing a billiard cue on the end of your finger. If it starts to tip (and because the weight is up near the top it is essentially unstable and *will* start to tip) the force applied by your finger must immediately change direction to push the base to one side and bring it back into line. This is done by having gyros of the guidance system immediately sense the change of attitude, and through the control system swivel the motor so that the thrust of the jet will point a little to one side and force the base in the required direction. In some of our missiles, the larger engine stays still, and small extra motors on the sides of the tail change direction to control the attitude. Since the thrust of these motors is much less than that of the main one, control is less positive, and the demand on the guidance system is correspondingly greater.

Now it takes considerable leverage to move a big rocket motor at full power, and if the motor is big enough and heavy enough the movement of the motor itself may tend to tip the missile still more and make the job still harder. Also, the bigger and heavier the motor to be moved, the harder it is to move it quickly and sensitively, and with every fraction of a second of delay in correction there is more risk that the missile will tilt over too far to be straightened out, or will turn so fast that the structure cannot stand the strain and will break up in flight.

If a cluster of motors is used, not all of them will have to be designed for swiveling. The center ring of engines can be fixed in position. At the same time, each of the outer motors can do more than just control the vertical attitude of the vehicle. By moving them all in the same radial direction, the vehicle can be caused to "roll" around its own long axis. Thus the same guidance system, applied through the same controls, can keep the vehicle on a steady path, correcting for error in all three axes—roll, pitch and yaw. Being smaller, each motor requires less force to move, and can move more quickly and delicately to do its job.

Finally, clustering motors permits the use of engines that have already had considerable flight testing in smaller missiles. A really new motor design from scratch ... is bound to have quite a few "bugs" in it. An exhaustive pre-flight test program on the ground can catch many of them, but not all. Inevitably there are a few defects that will not show up until the engine is sent out into space. This process is expensive and time consuming, and usually involves the loss of at least a few test vehicles before the engine can be considered operationally reliable. On the

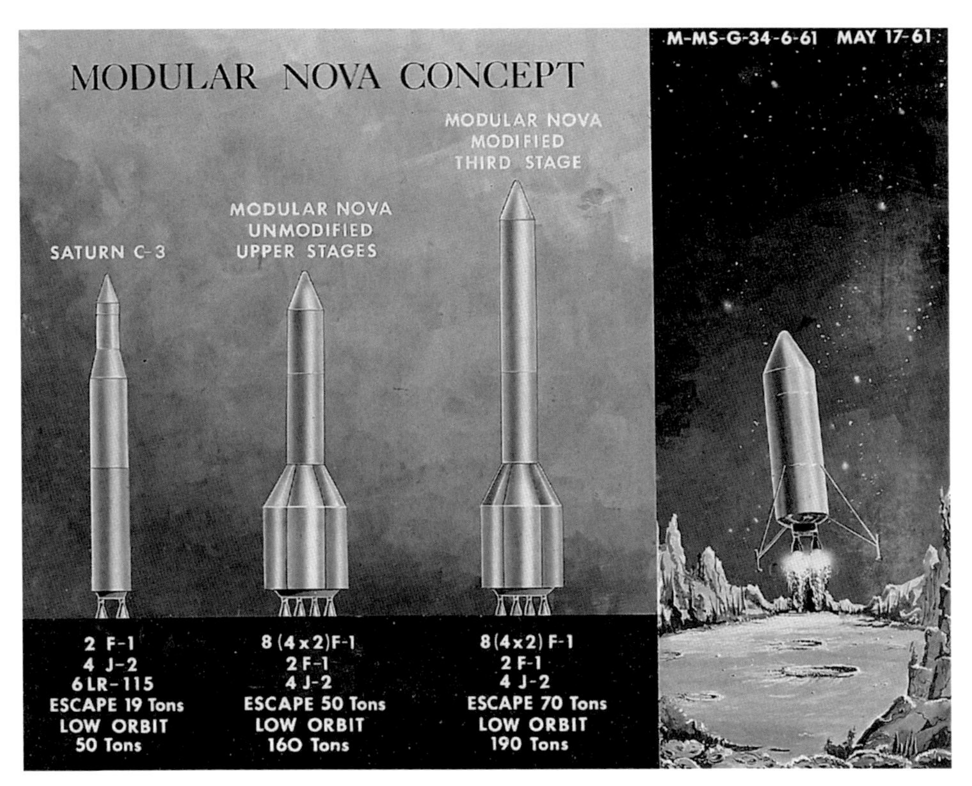


1-3 The Nova launch vehicle was conceived to perform a direct ascent to the Moon without employing spacecraft rendezvous, but it required a larger lunar landing craft. This illustration issued in May 1961 shows Saturn and Nova concepts using liquid propellants in all stages. (NASA/MSFC)

> other hand, the relatively minor changes that are required to adapt a proven motor to a new type of vehicle can be accomplished faster, and with substantially less risk. Thus, clustering of existing motors shortens the time scale and increases reliability.
>
> Tooling is also a factor in the time required to bring a new vehicle to the flight test phase. Using the single million-pound engine would mean the design of a whole new set of tanks and missile structure. We had tooling on hand for tanks of the Redstone diameter, and of the Jupiter size. Why not cluster tanks also? We would waste neither the time nor the money for new tank tooling, and we would not have to tackle again the problem of sloshing in a tank many times bigger than the Jupiter.

These were the theoretical advantages to clustering engines, but Medaris could draw on no experience of clustering to determine potential problems such as vehicle base heating, or the effect of one operating engine on the performance and exhaust flow of the neighboring engines. As the Atlas ballistic missile and the final version of the Navaho cruise missile both had clusters of only three engines, they could shed little light on how significant these issues would be in a cluster using eight engines.

Meanwhile, Congress had resolved that the nation should respond to the Soviet space threat by creating a special agency dedicated primarily to space exploration—a task which the NACA was not equipped to do. On July 29, 1958 President Dwight D. Eisenhower duly signed the National Aeronautics and Space Act. This effectively formed the National Aeronautics and Space Administration (NASA) which, when it was activated on October 1, 1958, would absorb the personnel and facilities of the NACA, and this older agency would cease to exist.

Further discussions between ABMA and ARPA during the summer succeeded in refining the proposed configuration of America's first super booster. ARPA Order No. 14-59 of August 15, 1958, directed ABMA and the Army Ordnance Missile Command (AOMC) to draw up a costed plan for the development and operation of the Juno V launch vehicle. Its performance had to be demonstrated by a first stage captive test no later than December 1959. The initial funding was $5 million. This was an amazingly compressed timeline. As there was no single engine available that could be clustered to deliver the requisite thrust, the Rocketdyne S-3D engine of the Jupiter missile, which currently produced 667 kN (150,000 pounds) of thrust, would require to be uprated. ABMA and the NACA then entered into discussions with the company about the modifications which would be essential and other modifications that would be advantageous, and to specify the development, testing and production schedules, and of course to agree the cost of the program. This actually involved two slightly different engines: the four outboard engines would have the capability to gimbal, and the four inboard engines would remain fixed. On September 11, 1958 a development contract was signed for this engine, which Rocketdyne designated the H-1. A Memorandum of Agreement between Maj. Gen. Medaris of AOMC and Roy Johnson of ARPA was signed on September 23 to expand the Juno V program from a demonstration of booster feasibility into "the development of a reliable high performance booster to serve as the first stage of a multistage carrier vehicle capable of performing advanced missions." The funding was increased to over $13 million. Additional funds were added to modify the Jupiter test stand in Huntsville, Alabama to accept the booster for captive testing. Finally, $7 million was allotted for launch facilities at the Atlantic Missile Range at Cape Canaveral, Florida.

Rocketdyne moved with such speed that by December 1958 it was able to make the first full power test of the H-1 engine. Work began to modify the test stand in Huntsville, and exploratory trips were made to Cape Canaveral to locate the launch complexes for the Juno V launch vehicle.

On January 27, 1959, after holding meetings with representatives from ARPA, ABMA and various internal groups, including the Space Task Group (STG), NASA issued *A National Space Vehicle Program*. It outlined the Saturn and Nova launch vehicles, the Vega upper stage proposed by the Jet Propulsion Laboratory (JPL), and the hydrogen–oxygen Centaur upper stage proposed by Lewis. The first stage of the Nova would be powered by four F-1 engines, its second stage by a single F-1 engine and its third stage by a cluster of hydrogen–oxygen engines. It would have additional upper stages as necessary. The Nova was to serve as a direct ascent launch vehicle to

1-4 As shown in this illustration issued in May 1961, one option for the Nova launch vehicle was to use solid propellant first and second stages, and an upper stage powered by four liquid propellant J-2 engines. (NASA/MSFC)

deliver a crew directly to the surface of the Moon without any form of rendezvous. Earlier in the month, NASA had signed a contract with Rocketdyne to proceed with the design, development and testing of the F-1, essentially taking over this program from the Air Force. This was the first time the Nova launcher was presented, but it would go through many design iterations powered by liquid, solid or a combination of both propellants. The Saturn launch vehicle was actually the Juno V, because von Braun's group had reasoned that Saturn was a more logical follow on to Jupiter. The change was made official by ARPA in February 1959, and a contract was issued to construct the Saturn blockhouse at the now-named Complex 34 at Cape Canaveral. As regards the launch vehicle which ABMA was already developing, which became the Saturn I, the configuration of the first stage powered by eight H-1 engines had been fixed, but its upper stages were very much under review. The options included adapting either the Atlas or Titan missiles! A more practical solution was to employ the Centaur, whose hydrogen–oxygen engines would deliver a significantly greater performance.

During this period, there were high level discussions regarding the proper role of ABMA and its personnel in space exploration. Instrumental in these discussions was

Dr. Herbert F. York, who was the Director of Defense Research and Engineering at the Department of Defense. In the fall of 1958, discussions began between Deputy Secretary of Defense Donald A. Quarles and NASA Administrator T. Keith Glennan about transferring to NASA both JPL and the obvious space exploration activities of ABMA—in the latter case essentially the rocket team led by Dr. Von Braun. Gen. Medaris understandably opposed such a transfer, but the prospect was enticing for von Braun since it would enable his group finally to pursue the exploration of space, which was what had kept most of the engineers and scientists together as a team for so many years. The situation for the Army was doubly distressing, because much of the work undertaken by JPL was done under contract to the Army. If the Army were to lose both groups, it would be unable to compete against the Air Force in the space arena. Whilst it was decided early on not to transfer JPL to NASA, the discussions concerning von Braun's group would run through into 1959.

## EARLY LUNAR STUDIES AND PROGRAMS

A secret program was initiated in March 1959 by the U.S. Army Chief of Research and Development, Lt. Gen. Arthur G. Trudeau. He directed the Chief of Ordnance to begin preliminary studies for setting up a manned military outpost on the Moon. The transmittal letter stated:

> The lunar outpost is required to develop and protect potential United States interests on the moon; to develop techniques in moon-based surveillance of the earth and space, in communications relay, and in operations on the surface of the moon; to serve as a base for exploration of the moon, for further exploration into space and for military operations on the moon if required; and to support scientific investigations on the moon.

U.S. intelligence estimates predicted that the Soviets would attempt to land men on the moon by 1967 in order to mark the 50th anniversary of the October Revolution. Trudeau's letter stated the degree of urgency bluntly:

> To be second to the Soviets in establishing an outpost on the moon would be disastrous to our nation's prestige, and in turn to our democratic philosophy.

This study was performed by the Preliminary Design Branch of ABMA, whose Chief, Heinz-Hermann Koelle, directed the team of researchers and engineers drawn from all Technical Services within the Army. Koelle was also the editor of the four-volume report that was issued in June 1959. Among the preliminary conclusions of the report was that Project Horizon represented "the earliest feasible capability for the U.S. to establish a lunar outpost. By its implementation, the United States can establish an operational lunar outpost by late 1966, with the initial manned landings [having] taken place in the spring of 1965."

The launch vehicles that formed the lifting capability to get crews and material for the lunar base to the Moon were Saturn I and Saturn II. In this configuration, the first stage of the Saturn I was to be powered by eight H-1 engines, each delivering a thrust of 837 kN; the second stage was a modified version of the Titan powered by a twin-chamber Aerojet LR-87 engine that burned RP-1 with LOX; and the third stage was the Centaur (already in development for the Atlas) powered by a pair of Pratt & Whitney RL-10 hydrogen–oxygen engines, each of which delivered 67 kN of thrust. It would be able to send 3.4 metric tons to the Moon. The first stage of the Saturn II would use uprated H-1 engines with a sea-level thrust of 1,116 kN, the second stage would be powered by hydrogen–oxygen engines and the final two stages would each use a single hydrogen–oxygen engine. Von Braun had argued to use RP-1 as the fuel for the upper stages, as in the case of the first stage, but this would have been a severe limiting factor in the launch vehicle's ability to send the desired payload to the Moon. After the Saturn II third stage had been refueled in low Earth orbit, it would be capable of sending almost 22.7 metric tons to the Moon, which was sufficient for a manned lander. The launch schedule would be so aggressive that Cape Canaveral would be insufficient to support Project Horizon. A total of eight launch pads would be required, and an equatorial launch complex either in Brazil or on Christmas Island was proposed in order to maximally exploit the velocity imparted to a launcher by the Earth's rotation. It was estimated that 61 Saturn I and 88 Saturn II launches would be required to fully configure the Earth-orbiting space station, the lunar base and its supporting infrastructure. Project Horizon was audacious, and its scope fairly yelled its staggering cost. This program even outstripped the Atlas and Titan ICBM programs managed by the Air Force. No doubt some members on the Project Horizon team felt that the program would never be realized. Nevertheless, the team recognized that high performance upper stages would be required to send manned missions to the Moon, and this was perhaps the greatest contribution to evolving launch vehicles for lunar exploration.

At the end of August 1959, Milton W. Rosen and F. Carl Schwenk of NASA presented a paper entitled *A Rocket for Manned Lunar Exploration* at the Tenth International Astronautical Congress in London. Rosen was Director of Launch Vehicles and Propulsion within the Office of Manned Space Flight. In the paper the authors described a five stage Nova launch vehicle for a direct ascent to the Moon. This approach necessitated landing a larger vehicle on the surface of the Moon in order to avoid the logistical complexities of spacecraft rendezvous—some people had doubts as to whether a rendezvous in space by two spacecraft was feasible at all. The first stage would be powered by a cluster of six F-1 engines, the second stage by two F-1 engine and the stages above would use hydrogen–oxygen engines.

During 1959 the Department of Defense continued to discuss with NASA the transfer of the Saturn program and the engineering personnel supporting it. Dr. York described this contentious transfer in his book:

> I had become Director of Defense Research and Engineering and thus acquired authority over all elements of the space program within the Department of Defense. I reviewed the whole space program [...] and made

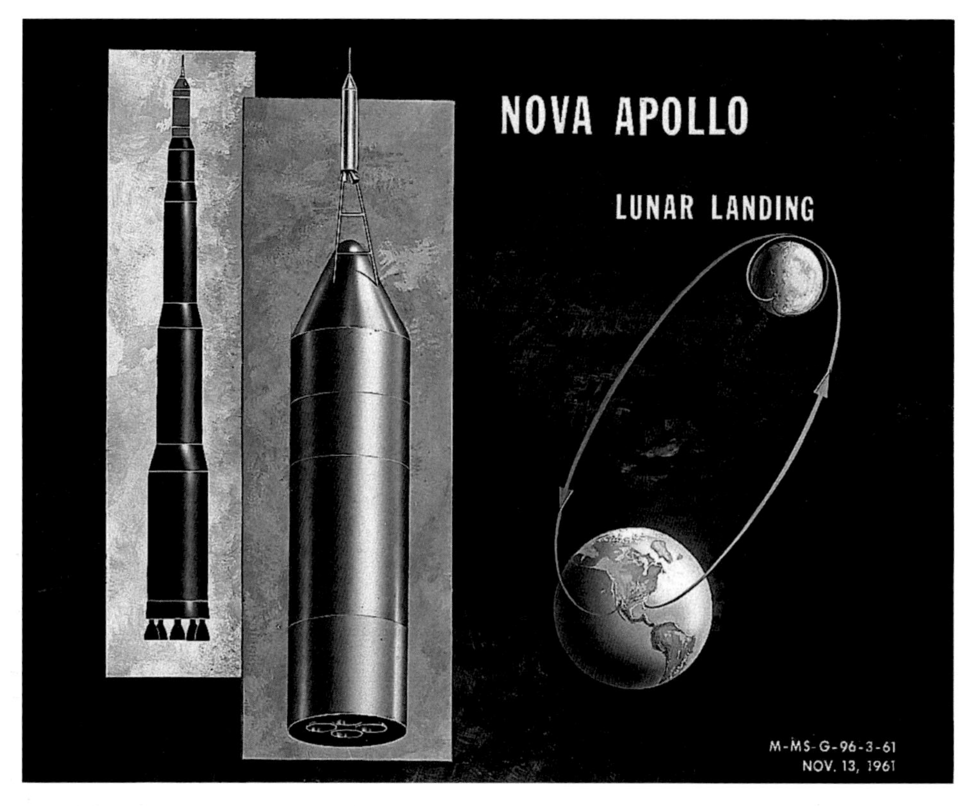


1-5 Direct ascent was NASA's preferred lunar mission mode in the first years of Apollo, and by November 1961 the liquid propellant Nova had evolved to this configuration. (NASA/MSFC)

two recommendations. The first was to Defense Secretary McElroy and was to the effect that the responsibility for developing all military-satellite launchers and for making all military-satellite launches should be transferred to the Air Force; that except for certain specifically named exceptions, all military-satellite-payload development be made the responsibility of the Air Force; and that the responsibilities and authorities of ARPA in military space programs be discontinued. McElroy accepted the recommendation, and a directive to this effect was issued late in the summer of 1959. My second recommendation was more far-reaching and had to go to the President and T. Keith Glennan, the Administrator of NASA, as well as to the Secretary of Defense for final implementation. In brief, at a meeting in the White House in late October, 1959, attended by the President, McElroy, Gates, Glennan, Dryden, General Nathan F. Twining, Chairman of the Joint Chiefs of Staff, and George B. Kistiakowsky, I recommended that (1) administrative responsibility for the Saturn booster be transferred from the Department of Defense (ARPA) to NASA, where the responsibility for manned space flight already was, and (2) that the Von Braun group be transferred to NASA along with the Saturn program. McElroy, Glennan, and the President accepted this recommendation, and transfers of authority and personnel were made a few months later.

By the end of December 1959, NASA's Saturn Vehicle Evaluation Committee, also known as the Silverstein Committee after its chairman Abe Silverstein, made its recommendations for the Saturn stages, with a research and a development schedule that included ten launches. Of the various stage configurations and their propellants, the committee identified six Saturn variations and recommended three, which were identified as Saturn C-1, C-2 and C-3. The first stages of the C-1 and C-2 were to be powered by eight H-1 engines. The C-3 would be powered by a cluster of engines producing a total of 2 million pounds of thrust—this later being specified as a pair of F-1 engines. The hydrogen–oxygen upper stages were listed as S-II, S-III, S-IV and S-V using the Pratt & Whitney LR-119, Pratt & Whitney RL-10 or Rocketdyne J-2 engines in clusters. This report would later prove instrumental in the early flight testing of the S-IVB stage of the Saturn V launch vehicle that would send a manned spacecraft to the Moon. In January 1960 the Saturn program was assigned a DX classification, which was the highest national priority, to place it on a par with the ICBM programs. Because the Saturn V was still several years away from realization, the recommendations of the Saturn Vehicle Evaluation Committee concerned only the Saturn I. At a small luncheon in Washington, D.C., Abe Silverstein suggested the name Apollo for the follow-on program after Mercury, but the name was adopted for the lunar landing missions. Gemini would later emerge as the intermediate manned program. That month, a mockup of the Saturn I first stage was installed on the test stand in Huntsville in order to verify how it would mate with the stand and servicing hookups. This was replaced in February by the SA-T, the first Saturn test stage. In March, two of the eight H-1 engines of the SA-T passed an eight second static firing test. A four engine test was made early in April. The first eight engine test firing occurred on April 29, 1960. This Saturn booster continued with a methodical testing program. After evaluating bidders' proposals for the second stage, which was in fact the S-IV, NASA awarded the contract to the Douglas Aircraft Company. In May Rocketdyne was selected to develop the J-2 hydrogen–oxygen engine for the upper stages. The series of Saturn I launch vehicles were designated SA-1 through SA-10. The first stage for SA-1 was assembled in Huntsville, and it successfully completed a series of tests that culminated on June 15, 1960 in a 122-second firing of all eight H-1 engines.

On March 15, 1960, President Eisenhower signed an Executive Order formally renaming the AOMC facilities in Huntsville as the George C. Marshall Space Flight Center (MSFC). This officially began operations on July 1, with Dr. von Braun as its first Director. The transfer of personnel working on Saturn from the Army to NASA would occur over a period of months in a manner designed to minimize disruption of existing Army programs. NASA established the Office of Launch Vehicle Programs at MSFC. In June, NASA had created the Launch Operations Directorate with Dr. Kurt H. Debus as its Director. The first week in July, the U.S. House Committee on Science and Astronautics stated, "A high priority program should be undertaken to place a manned expedition on the moon in this decade. A firm plan with this goal in view should be drawn up and submitted to the Congress by NASA." The Committee also recommended that the F-1 engine program should be accelerated in anticipation of the finalization of the Nova launch vehicle design. In a memorandum on July 25, Abe Silverstein, NASA's Director of Space Flight Programs, informed Harry Goett, the Director of the Goddard Space Flight Center, that the NASA Administrator had approved the name Apollo for the advanced manned space program.

On July 28–29, 1960, a NASA-Industry Program Plans Conference was held in Washington, D.C. The agency presented its mission plans, including Project Apollo, and the status of its launch vehicles, infrastructure, research and other related areas. It also announced a series of follow-up conferences to be hosted by various NASA centers during the remainder of the year at which aerospace industry representatives would receive more detailed briefings. At the conference at the Langley Research Center in Hampton, Virginia in September, NASA conducted a bidders' briefing and issued a Request for Proposals (RFP) for an advanced manned spacecraft to fly the Apollo missions. Interested companies had only 30 days to prepare their proposals. On October 10 the agency began to evaluate the proposals, and on October 25 issued three feasibility study contracts to General Electric of Philadelphia, Pennsylvania, Convair/Astronautics of San Diego, California, and the Martin Company of Baltimore, Maryland. Meanwhile, geopolitical events would firm up the objectives of the manned space program.

## KENNEDY'S LUNAR DECISION

In the November 1960 presidential election, senator John F. Kennedy narrowly beat Richard M. Nixon by 100,000 out of a total of over 68 million popular votes, but it was Kennedy's slim lead in the electoral college votes that won him the election. He was sworn in as the 35th President on January 29, 1961, with Texas Senator Lyndon B. Johnson as Vice President. Earlier that month, the Space Exploration Program Council at NASA Headquarters listened to presentations on a manned lunar landing by direct ascent, Earth-orbit rendezvous or lunar orbit rendezvous. This established the Manned Lunar Exploration Working Group to identify the elements which each approach would require to achieve a manned lunar landing. Chaired by George M. Low this group included Ernest Pearson, Eldon Hall, Oran Nicks and A.M. Mayo of NASA Headquarters, Heinz-Hermann Koelle of the Marshall Space Flight Center, and Maxime A. Faget of the Space Task Group at Langley. Low's Committee spent a month evaluating mission modes and possible launch vehicles, and on February 7 it presented *A Plan for Manned Lunar Landing* to Associate Administrator Robert C. Seamans at NASA Headquarters. It proposed either using the Nova for direct ascent or the Saturn C-2 to assemble the spacecraft in Earth-orbit and to accumulate sufficient propellant to send it to the Moon. In the view of the committee, the riskier approach would be to rely upon the development of the large Nova launch vehicle.

In February 1961, James E. Webb was named as the new NASA Administrator. On April 12, the Soviet Union again shocked the world with the successful launch, orbiting and landing of cosmonaut Yuri Gagarin. Once again, the Soviet Union was

1-6 Dr. Wernher von Braun, the first Director of the Marshall Space Flight Center, was skilled at presentation and an articulate, persuasive negotiator. Here during a visit to NASA Headquarters in November 1961 he presents an idea to Brainerd Holmes (left), Director of the Office of Manned Space Flight, and Dr. Nicholas Golovin (center), Director of Systems Engineering. (NASA/MSFC)

the leader in space exploration and the United States was trailing behind. This event not only confirmed that the Soviets had impressive boosters, but also displayed that country's technological capability to keep a human alive in space and safely return him to Earth.

On April 20—immediately after the failed Bay of Pigs invasion of Cuba designed to overthrow the communist dictator Fidel Castro and his government—Kennedy issued a memorandum to Lyndon Johnson, who was Chairman of the Space Council, posing five questions to determine "... where we stand in space." They were:

1. Do we have a chance of beating the Soviets by putting a laboratory in space, or by a trip around the moon, or by a rocket to land on the moon, or by a rocket to go to the moon and back with a man. Is there any other space program which promises dramatic results in which we could win?
2. How much additional would it cost?
3. Are we working 24 hours a day on existing programs. If not, why not? If not, will you make recommendations to me as to how work can be speeded up.
4. In building large boosters should we put out emphasis on nuclear, chemical or liquid fuel, or a combination of these three?
5. Are we making maximum effort? Are we achieving necessary results?

To get a realistic appraisal of the situation, Johnson consulted the Secretary and Deputy Secretary of Defense, Gen. Bernard Schriever of the Air Force, Adm. Hayward of the Navy, Dr. Von Braun, NASA Administrator James Webb and Deputy Administrator Hugh Dryden, representatives from the Bureau of the Budget, and others from private industry. One week later, Johnson responded with a six page memorandum. Among his conclusions were the following:

> The U.S. can, if it will, firm up its objectives and employ its resources with a reasonable chance of attaining world leadership in space during this decade. This will be difficult but can be made probable even recognizing the head start of the Soviets and the likelihood that they will continue to move forward with impressive successes. In certain areas, such as communications, navigation, weather, and mapping, the U.S. can and should exploit its existing advance position.
>
> Manned exploration of the moon, for example, is not only an achievement with great propaganda value, but it is essential as an objective whether or not we are first in its accomplishment—and we may be able to be first.

This memorandum went on to answer Kennedy's specific questions. Other than what the Soviets had demonstrated to date, it was hard to guess their intentions or future booster capabilities, but the memorandum stated that by making a maximum effort the U.S. might be able to achieve a manned circumlunar mission or a manned lunar landing and return by 1966 or 1967. It estimated the cost of achieving this at roughly $1 billion per year during a period of ten years. The Department of Defense urged the development of a large solid-propellant booster. With regard to Kennedy's third question, the development of the Saturn I booster and Centaur upper stage were limited by funds and a low-key manpower work schedule. "This work can be sped up through a firm decision to go ahead faster if accompanied by additional funds needed for the acceleration," the document stated. It also recommended liquid, solid and even nuclear propulsion should all be accelerated. Regarding Kennedy's final question, the answer was telling and blunt: "We are neither making maximum effort nor achieving results necessary if this country is to reach a position of leadership."

On April 29, von Braun forwarded his detailed assessment to the Vice President, making two honest assessments of manned lunar missions:

> We have a sporting chance of sending a 3-man crew around the moon ahead of the Soviets (1965/1966). However, the Soviets could conduct a round-the-moon voyage earlier if they are ready to waive certain emergency safety features and limit the voyage to one man. My estimate is that they could perform this simplified task in 1962 or 1963.
>
> We have an excellent chance of beating the Soviets to the first landing of a crew on the moon (including return capability, of course). The reason is that a performance jump by a factor of 10 over their present rockets is necessary to accomplish this feat. While today we do not have such a rocket, it is unlikely that the Soviets have it. Therefore, we would not have to enter the race toward this obvious new goal in space exploration against hopeless odds favoring the Soviets.

1-7 The Saturn I booster proved the feasibility of clustering large liquid propellant engines. This program also set many precedents for NASA and the Marshall Space Flight Center in constructing, testing and launching a multi-engine boosters which were later applied to the Saturn V. (NASA/MSFC)

> With an all-out crash program I think we could accomplish this objective in 1967/68.

He answered Kennedy's final question with equal frankness, and then elaborated on what should be done:

> No, I do *not* think we are making maximum effort. In my opinion, the most effective steps to improve our national stature in the space field, and to speed things up would be to
>
> - Identify a few (the fewer the better) goals in our space program as objectives of the highest national priority (For example: Let's land a man on the moon in 1967 or 1968.)
> - Identify those elements of our present space program that would qualify as immediate contributions to this objective. (For example, soft landings of suitable instrumentation on the moon to determine the environmental conditions man will find there.)

- Put all other elements of our national space program on the "back burner."
- Add another more powerful liquid fuel booster to our national launch vehicle program. The design parameters of this booster should allow a certain flexibility for desired program reorientation as more experience is gathered.

  Example: Develop in addition to what is being done today, a first-stage liquid fuel booster of twice the total impulse of Saturn's first stage, designed to be used in clusters if needed. With this booster we could

  a. double Saturn's presently envisioned payload. This additional payload capability would be very helpful for soft instrument landings on the moon, for circumlunar flights and for the final objective of a manned landing on the moon (if a few years from now the route via orbital re-fueling should turn out to be the more promising one.)

  b. assemble a much larger unit by strapping three or four boosters together into a cluster. This approach would be taken should, a few years hence, orbital rendezvous and refueling run into difficulties and the "direct route" for the manned lunar landing thus appears more promising.

  Summing up, I should like to say that in the space race we are competing with a determined opponent whose peacetime economy is on a wartime footing. Most of our procedures are designed for orderly, peacetime conditions. I do not believe that we can win this race unless we take at least some measures which thus far have been considered acceptable only in times of a national emergency.

Yours respectfully,
Wernher von Braun

May 1961 was a hectic one for lunar mission proposals, planning and legislative activity. It would also see the launch of America's first astronaut. On May 2, Robert C. Seamans, NASA's Associate Administrator, established the Ad Hoc Task Group for a Manned Lunar Landing Study. Chaired by William A. Fleming, it would make recommendations regarding all aspects surrounding manned lunar missions—with special emphasis on whether to use a Saturn or Nova launch vehicle. It was to report by mid-June.

On May 4, Vice President Johnson received a memorandum from Congressman Overton Brooks, Chairman of the House Committee on Science and Astronautics. Brooks was very aware of the technological and political beating the U.S. had taken in the face of Soviet space exploration successes, and his opening paragraphs clearly stated what the nation must do, and that the Congress would undertake whatever was necessary in order for America to gain leadership in space.

> It is my belief, and I think on this point that I can speak for our committee, that the United States must do whatever is necessary to gain unequivocal leadership in Space Exploration.
>
> This means the procurement and utilization of sufficient scientific talent, labor and material resources as well as the expenditure of sufficient funds. This means

> working around the clock, if need be, in all areas of our Space program—not just a few.
>
> The reason is patent. Rightly or wrongly, leadership in space research and exploration has assumed such a powerful position among the elements which form the political stature of our country in the eyes of the world that we cannot afford to slight it in any fashion whatsoever. This is perhaps even truer of the non-military phase of our national space endeavor than it is of the military. Obviously, neither phase can be slighted.

Brooks pointed out that the Soviet Union was spending at least two percent of its gross national product on its space exploration activities, whereas the United States was spending roughly one half of one percent. He spoke for the entire Committee when he wrote: "We believe that a particular effort must be made to strengthen such programs as Apollo, Saturn, Rover (nuclear powered rocket engine) and the solid-segmented and F-1 liquid engine concepts." He stated the Committee's position on lunar exploration: "We should pursue vigorously our man-in-space program. We cannot concede the Moon to the Soviets, or it is conceivable that the nation which controls the Moon may well control the Earth."

Even as Vice President Johnson studied Brooks' memorandum, national attention was focused on a Redstone rocket at Cape Canaveral bearing the Mercury capsule *Freedom 7*. Astronaut Alan B. Shepard was preparing to become the first American to enter space. Early in the morning of May 5, Shepard entered his cramped capsule and the hatch was bolted into place. It was vital that America's first manned space flight be a success. As the Redstone lifted off at 9:34 AM, the hearts of millions of Americans skipped a beat. The rocket released the capsule on a trajectory that had its apogee at an altitude of 187 kilometers. As the capsule coasted, Shepard was able to experience weightlessness, but in the cramped cabin he had to remain strapped in his couch. As this was a suborbital flight, soon the capsule's automated system had turned it around to face the heat shield forward ready for the fiery dive back into the atmosphere. It splashed into the Atlantic after a flight lasting less than 16 minutes, but in that time Americans had felt triumphant. The mission was a complete success, and the U.S. manned space program had officially begun.

President Kennedy spent the next several weeks refining a special speech on the theme of "Urgent National Needs," which he delivered to a joint session of Congress on May 25. Under the section on Space, he said:

> Finally, if we are to win the battle that is now going on around the world between freedom and tyranny, the dramatic achievements in space which occurred in recent weeks should have made clear to us all, as did the Sputnik in 1957, the impact of this adventure on the minds of men everywhere who are attempting to make a determination of which road they should take. Since early in my term, our efforts in space have been under review. With the advice of the Vice President, who is Chairman of the National Space Council, we have examined where we are strong and where we are not, where we may succeed and where we may not. Now it is time to take longer strides, time for a great new American enterprise, time for this nation to take a clearly leading role in space achievement, which in many ways

may hold the key to our future on earth.

I believe we possess all the resources and talents necessary. But the facts of the matter are that we have never made the national decisions or marshaled the national resources required for such leadership. We have never specified long-range goals on an urgent time schedule, or managed our resources and our time so as to insure their fulfillment.

Recognizing the head start obtained by the Soviets with their large rocket engines, which gives them many months of lead-time, recognizing the likelihood that they will exploit this lead for some time to come in still more impressive successes, we nevertheless are required to make new efforts on our own. For while we cannot guarantee that we shall one day be first, we can guarantee that any failure to make this effort will make us last. We take an additional risk by making it in full view of the world, but as shown by the feat of astronaut Shepard, this very risk enhances our stature when we are successful. But this is not merely a race. Space is open to us now. and our eagerness to share its meaning is not governed by the efforts of others. We go into space because whatever mankind must undertake, free men must fully share.

1-8 The Saturn I's four H-1 inboard engines were fixed in position and tightly clustered, while the four outboard H-1 engines could gimbal to steer the large booster. (NASA/MSFC)

I therefore ask the Congress, above and beyond the increases I have earlier requested for space activities, to provide the funds which are needed to meet the following national goals:

First, I believe that this nation should commit itself to achieving the goal, before this decade is out, of landing a man on the moon and returning him safely to the earth. No single space project in this period will be more impressive to mankind, or more important for the long-range exploration of space; and none will be so difficult or expensive to accomplish. We propose to accelerate the development of the appropriate lunar space craft. We propose to develop alternate liquid and solid fuel boosters, much larger than any now being developed, until certain which is superior. We propose additional funds for other engine development and for unmanned explorations—explorations which are particularly important for one purpose which this nation will never overlook: the survival of the man who first makes this daring flight. But in a very real sense, it will not be one man going to the moon—if we make this judgment affirmatively, it will be an entire nation. For all of us must work to put him there.

## LUNAR LAUNCH VEHICLE STUDIES

In his speech, Kennedy went on to list the funding requirements for other aspects of space exploration, and the members of Congress who heard the numbers did not flinch. That same day, Robert C. Seamans requested that the Directors of the Office of Launch Vehicle Programs and the Office of Advanced Research Programs form a committee made up of individuals within their respective groups and other selected individuals within NASA to study the launch vehicles and mission methods needed to accomplish a lunar landing. Bruce T. Lundin was appointed as chairman. As part of this process, John C. Houbolt presented the *Lunar Orbit Rendezvous Plan* devised by a team at Langley. In fact, LOR was the underdog in terms of 'mission mode', and Houbolt would spend nearly two years working to convince the powers-that-be at NASA of its advantages. The Space Task Group established the Flight Vehicles Integration Branch. Its membership included Maxime A. Faget, Chief of the Flight Systems Division. On June 10, the Lundin Committee presented its conclusions in *A Survey of Various Vehicle Systems for the Manned Lunar Landing Mission*. It had studied the Saturn C-2 and C-3, and recommended using two or three Saturn C-3s to assemble in Earth orbit the spacecraft required to take a crew to the Moon. A week later, the Fleming Committee issued its report. It said that a manned mission to the Moon should be feasible within the decade, and that the pacing items would be the development of the first stage of the launch vehicle and associated testing facilities.

On June 20, 1961, Seamans established yet another Ad Hoc Task Group, in this case under Donald H. Heaton with the participation of the directors of the Offices of Launch Vehicles Programs, Space Flight Programs, Advanced Research Programs and Life Sciences Programs. This was to evaluate Fleming's and Lundin's findings, establish the necessary resources and plans to implement a manned lunar landing by 1967 using the Saturn C-3 and the appropriate rendezvous procedures. On June 23, Dr. von Braun said that engineering work on the Saturn C-2 would be discontinued to focus on the Saturn C-3 and Nova concepts. Meanwhile, on 5 June a dedication ceremony had been held to mark the completion of work at Launch Complex 34. On July 25, the Space Task Group issued Working Paper 1023 entitled *Project Apollo: A Description of A Saturn C-3 and NOVA Vehicle*. As described by the STG, the Saturn C-3 had a first stage powered by a pair of F-1 engines, the second stage had a cluster of four J-2 hydrogen–oxygen engines, and the third stage had a cluster of six RL-10A-3 hydrogen–oxygen engines. It would be capable of placing 45 metric tons into low earth orbit, or sending 17.7 metric tons to the Moon. The launch facilities, performance characteristics, staging sequence events, trajectory and miscellaneous other information was provided. The Apollo crew capsule, launch escape system and service module illustrated in this paper were remarkably close to the designs which eventually emerged. The Nova was, the authors admitted, "in the very preliminary stages of planning and design." As described, the Nova was a beast. It first stage had a diameter of 13.2 meters and a height of 103 meters, and was powered by a cluster of eight F-1 engines with an overall thrust of 53,980 kN. The second stage would be powered by eight J-2 engines with a total thrust of 8,264.8 kN. The third stage had a pair of J-2s. The Nova was conceived to support direct ascent, without the need for a rendezvous either in Earth orbit or lunar orbit. As the spacecraft designed for direct ascent would be much larger than for a mission involving a specialized lunar lander, the Nova could theoretically send 82 metric tons to the Moon. However, because the Nova was only at a preliminary state of development, the Space Task Group had less to say about it than it had regarding the Saturn C-3. Nevertheless, the sheer scale of the Nova would significantly influence the proposed design of the 'spaceport'.

A key meeting was held in Dr. von Braun's office on August 7, 1961 to discuss MSFC's position on advanced launch vehicles: specifically the Saturn C-3, C-4 and Nova. Those present included Dr. Eberhard F.M. Rees (Deputy Director for Research and Development), Mr. Erich W. Neubert (Director, Systems Analysis and Reliability at ABMA prior to becoming Associate Deputy to Rees), Heinz-Hermann Koelle (Director, Future Projects Office), Dr. William A. Mrazek (Director, Propulsion and Vehicle Engineering Division), Hans H. Maus (Director, Central Planning Office), Dr. Oswald H. Lange (Director, Saturn Systems Office), Dr. Kurt H. Debus (Director, Launch Operations), Dr. J. C. McCall (Assistant to Dr. von Braun), Wilson B. Schramm of MSFC and Edward Morris.

Mr. Schramm reported that the Space Task Group recommended the development of a 160 inch diameter (4.0 meter) solid propellant motor, and then he delivered a report by JPL on a proposed launch vehicle using only solid propellants. Mr. Koelle presented a document by the Future Projects Office on the Saturn C-3 with two F-1 engines in its booster, the C-4 with four such engines and the Nova with eight. Dr. von Braun discussed manned lunar missions using various launchers, then asked for each person's recommendations regarding the best and quickest means of achieving a manned lunar landing. Mr. Schramm favored rapid development of the Nova for direct ascent, with the Saturn C-3 as a backup. Mr. Koelle recommended Earth orbit rendezvous using the C-1 launch vehicle, development of the C-4, and use of high energy propellants for the Nova upper stages. Mr. Maus recommended development

1-9 Dr. Wernher von Braun welcomes President John F. Kennedy to the Marshall Space Flight Center in September 1962. (NASA/MSFC)

of the C-4, and felt that Congress would balk at funding the Nova. Dr. Lange urged upgrading the Saturn I to employ the J-2 engine when this became available, and to develop the Saturn C-3. Mr. Mrazek pointed out that the base heating problem of a C-4 configuration was more serious than on the C-2, recommended development of the C-3 immediately, and the Nova with a first stage having a cluster of fuel tanks or solid rockets. Mr. Morris felt that it would be impractical to fund a Saturn C-3 and Nova simultaneously, and urged development of the C-4. Dr. Rees was the voice of pessimism, opining that the U.S. did not have a chance of beating the Soviets to the Moon at the current pace and funding, and that the Soviets would probably establish a base on the Moon so as to deny the Americans access. Nevertheless, he concluded that the U.S. should embark on a crash program with appropriate funding to be first to achieve a lunar landing. Dr. Debus recommended a solid rocket booster for the Saturn C-1, favored development of solid and liquid rocket boosters simultaneously but favored solid rocket development due to their greater simplicity, urged development of the Saturn C-3 without delay, and recommended Nova at a later date after further study of its various possible configurations. Mr. Neubert said NASA should limit itself to the development of liquid propellant systems in order to achieve a lunar landing by 1967, and urged development of the Saturn C-4 because this would require no new manufacturing facilities, although further study would be required regarding how its stages should be transported. After listening to each individual's recommendations, Dr. von Braun gave his own recommendations. He felt the Saturn C-4 would permit the greatest payload to be delivered to the Moon, but added that since the C-4 could not guarantee direct ascent it would be wise to include Earth orbit rendezvous.

The consensus was that the Saturn C-3 should be developed as soon as possible, and the Nova should be studied for several more months prior to making a final decision as to the best launch vehicle for the lunar landing mission. The Nova had to remain an option until the decision was made regarding how to fly to the Moon: if this were to be by direct ascent then the Nova would be required, but if some form of rendezvous was decided upon then a smaller rocket would suffice.

The most detailed studies for the Nova class conceived for direct ascent came out of the Future Project Office at MSFC. Koelle stated in April 1961, first there was Nova-A, Nova-B and Nova-C. Nova-A would employ two F-1 engines with a total thrust of 13,495 kN, or roughly twice the thrust of the Saturn C-1. Nova-B had 2.5 times the thrust of Nova-A. The first stage of the Nova-C would have eight F-1 engines delivering a total of 53,980 kN of thrust at sea level. Its second stage would have two F-1s, the third stage four J-2s, the fourth stage six RL-10s, and the fifth stage two RL-10s. The Future Project Office also studied solid propellant first and second stages and hydrogen–oxygen upper stages. Illustrations of the Nova drawn up at this time showed an ungainly clustering of the first stage solid rocket motors, with single upper stages towering above. The designs were refined throughout 1961 and into 1962. The history of rocket boosters to date had indicated that larger boosters were always desirable, and it was conceivable that payloads might be developed that would require something even more powerful than the 'advanced' versions of the Saturn, in which case it might be wise to develop Nova even if it was decided to use rendezvous for the lunar mission. But the nation's priority in space was a mission to land a man on the Moon, and the issue of launchers would have to remain open until the decision was made regarding how this would be done.

## REFINING THE LUNAR MISSION MODE

During the studies and proposals for launch vehicle drawn up in the late 1950s and the early 1960s, NASA deliberately refrained from making any decision that would effectively lock in the design for the rocket that would send astronauts to the Moon. Earth orbit rendezvous would require numerous Saturn launches, rendezvous at an orbiting station to assemble the stages which would travel to the Moon and facilitate the return to Earth. Direct ascent using a Nova would avoid this complex rendezvous scenario but at the expense of developing a launch vehicle of unprecedented size. In either case, the entire spacecraft would land on the Moon. A third approach began to emerge in 1959 when Eugene Draley, Associate Director of the Langley Research Center, set up a working group to study the issues concerning lunar exploration. The membership included Paul Hill of the Pilotless Aircraft Research Division, David Adamson of the Supersonic Aerodynamics Division, John C. Houbolt of the Dynamic Loads Division, Bill Michael of the Theoretical Mechanics Division, Albert Schy of the Stability Research Division and Samuel Katzoff of the Full-Scale

Research Division. The group was directed by Clinton E. Brown of the Theoretical Mechanics Division. In the summer of 1959 Dr. Houbolt formed two subgroups to study orbital mechanics and the concept of rendezvous in space. Houbolt became a strong proponent of rendezvous as an essential aspect of any manned space program, irrespective of whether this involved stations in Earth orbit or the exploration of the Moon and the planets of our solar system. In particular, he felt that lunar orbit rendezvous was the key to the cheapest means of achieving lunar exploration. In essence, a rocket would send to the Moon a crew capsule that incorporated its own propulsion system and ferried a separate lunar landing craft having a descent stage and an ascent stage. Once in lunar orbit, the crew would transfer to the landing craft, separate from the mothership and land on the Moon. After the surface exploration was complete, the ascent stage would lift off, leaving the descent stage behind. Once the ascent stage had rendezvoused and docked with the mothership, the ascent stage would be jettisoned. The spacecraft would leave lunar orbit. On approaching Earth, the propulsion system would be jettisoned, leaving the crew capsule to reenter the atmosphere. The beauty of this lunar mission mode was that the lunar landing craft would be of the lightest possible configuration. This reduced the mass of the payload that the booster had to lift to such a degree that the Nova was not necessary. In fact, if the spacecraft could be launched in an integrated configuration there would be no need for multiple launches and Earth orbit rendezvous.

Houbolt promoted the advantages of the LOR mode throughout NASA during 1960 and 1961. In fact, it became a crusade. In contrast to the other two modes, he felt that it was the most efficient, indeed elegant, means of achieving a manned lunar landing and return. But the reception he received varied from disinterest to outright hostility. Many senior NASA people preferred direct ascent, whilst others, such as Dr. von Braun, leaned toward Earth orbit rendezvous. If a rendezvous in Earth orbit were to fail, then the astronauts would be able to return to Earth. The fear was that if the rendezvous in lunar orbit were to fail then the crew would be stranded. But there were risks to the astronauts regardless of the lunar mode selected, running from the moment of launch to the moment of landing back on Earth. As Houbolt explained in an interview years after the Apollo program had been successful executed, there was "virtually universal opposition—no one would accept it—they would not even study it." The Low Committee, the Lundin Committee and the Heaton Committee all rated LOR as the method least likely to succeed. But Houbolt persevered. In August 1961 he presented LOR to the Golovin Committee, and this time was well received. Even those in the Space Task Group had begun to consider it seriously. What had changed their minds was the difficulty of landing a single large spacecraft on the Moon; if the ship were to be long and thin and had legs around its base, it would be quite likely to tip over if it came down on uneven ground, yet how would it be landed in a manner other than upright? A specialized vehicle would be easier to land. Houbolt sent an extensive letter directly to NASA Administrator Seamans describing LOR at length. Seamans answered assuring Houbolt that his proposal would get a fair evaluation at Headquarters. Seamans passed the letter on to D. Brainerd Holmes, the new head of the Office of Manned Space Flight (replacing Abe Silverstein). Houbolt's letter was also reviewed by Dr. George Low, Director of Spacecraft and Flight Missions. LOR was finally getting serious consideration, but it would be months yet before a lunar mode decision would be made.

## THE SATURNS EMERGE

In September 1961, NASA announced that the Michoud Ordnance Plant near New Orleans, Louisiana, a war-surplus government facility, had been selected as the site for fabrication, assembly and checkout of the Saturn C-3 and larger proposed launch vehicles. On September 17, NASA issued an RFP to more than 30 companies for the engineering and manufacture of the first stage of the Saturn I launch vehicle, and a bidders' conference was held on September 26. This contract was eventually issued to the Chrysler Corporation.

In the first week of October, JPL published an addendum to its original Technical Memorandum No. 33-52; the original document evaluated direct ascent using a solid

1-10 Dr. John C. Houbolt of the Langley Research Center in Virginia championed the lunar orbit rendezvous mission mode for Apollo. NASA's decision to use this scheme shelved plans for the Nova launch vehicle. (NASA/Langley)

fuel Nova and a liquid propellant Saturn C-3, and a fourth mode described as lunar surface rendezvous whereby the astronauts would assemble the vehicle they were to use in returning to Earth! The addendum elaborated on the long term implications of conducting each launch vehicle program individually or in parallel. One astounding figure was the estimated mass of the solid fuel Nova at 11,340 metric tons! The solid propellant Nova would probably be able to be developed faster than the liquid propellant Nova. However, parallel development of both the solid and liquid Nova would be financially prohibitive. The report stated that either version of the Nova should be employed of manned missions to Mars.

Meanwhile, preparations continued at Cape Canaveral to launch the first Saturn. SA-1 was a test on many levels: not just of its performance in flight but also of the entire chain from manufacture to shipping the launch vehicle to the Cape, assembly on the pad, systems checkout, telemetry and tracking. The first four launch vehicles, known as Block I, would incorporate dummy upper stages. The fifth test would have a 'live' second stage. The third stage (the S-V) had been canceled.

On October 27, 1961 SA-1, weighing 417 metric tons, lifted off from Launch Complex 34. All eight H-1 engines burned perfectly, with the vehicle reaching an apogee of 136.8 km on a suborbital trajectory. The inboard engines shut down at their programmed 109 seconds, and the outboard engines at 115 seconds. The entire vehicle fell into the Atlantic within kilometers of the projected area. Dr. von Braun's team from MSFC and Dr. Debus' team at the Cape were jubilant at this milestone in the program to land a man on the Moon.

On November 10, 1961, NASA received proposals from five companies to make the first stage of the 'advanced' Saturn, in whatever configuration was decided. That month the rapidly expanding Space Task Group, which had outgrown Langley, was relocated to the recently built Manned Spacecraft Center (MSC) in Houston, Texas. Dr. Robert R. Gilruth, the head of the Space Task Group, became the Director of the new Center. On November 20, Milton W. Rosen, Director of Launch Vehicles and Propulsion, sent D. Brainerd Holmes the results of his working group's investigation into large launch vehicles. Among the recommendations were:

- To immediately pursue a program to develop a rendezvous capability.
- To develop a launch vehicle that had five F-1 engines powering the booster, either four or five J-2 engines powering the second stage, and a single J-2 on the third stage. This configuration was initially designated the Saturn C-5.
- Direct ascent should be emphasized for achieving the first manned landing in the soonest time possible. This would require a Nova with eight F-1 engines in the booster, four M-1 engines in the second stage, and one J-2 engine in the third stage. It should be developed as a top national priority.
- Large solid fuel boosters should not be considered for the lunar landing program unless they were developed with other national requirements in mind.
- Development of the S-IVB with one J-2 engine should begin with test flights as an upper stage to the Saturn IB in 1964. It was recommended that this be used as the third stage of the Saturn C-5 and Nova.

- Although NASA had no current requirement for the Titan III proposed by the Air Force, if this rocket was developed NASA should maintain contact with the Air Force with an eye on its potential use by NASA.

On December 7, 1961, NASA chose the Boeing Company as a potential builder of the S-IC first stage of the Saturn C-5. In a significant move, D. Brainerd Holmes sent a memorandum to Robert C. Seamans with a preliminary development plan for a manned space program to follow Mercury, named Mercury Mark II, the primary goal of which was to develop and perfect the rendezvous and docking capability even if direct ascent were chosen, in order to facilitate future manned space programs involving space stations. The next day, Robert R. Gilruth announced that a two-manned spacecraft, both larger and heavier than the Mercury capsule, would be developed by the McDonnell Aircraft Corporation; the company that had built the Mercury spacecraft. It would be launched by a two-stage Titan II supplied by the Air Force. In January, NASA announced that the name of this program would be Gemini.

On December 21, D. Brainerd Holmes announced the formation of the Manned Space Flight Management Council. It included key NASA personnel: Dr. Robert R. Gilruth, Dr. Wernher von Braun, Milton W. Rosen, Dr. George M. Low, Dr. Eberhard F.M. Rees, Dr. Charles H. Roadman (Director of Aerospace Medicine) and Dr. Joseph F. Shea (Holmes' deputy). At its first meeting, the Council decided that, in line with the recommendations of Rosen's study, the first stage of the Saturn C-5 should be powered by five F-1 engines, the second stage by five J-2 engines and the third stage by one J-2 engine. It also urged Boeing to commence complete design studies and a detailed manufacturing and development plan for the S-IC. On January 5, 1962, North America Aviation issued drawings of the three-man Apollo capsule. Then, on January 9, NASA formally announced that the Saturn C-5 had been chosen as the launcher for the manned lunar landing mission. Gilruth telephoned Houbolt and asked him to make a presentation on orbital rendezvous to the Manned Space Flight Management Council. Houbolt did so, along with fellow Langley engineer Charles W. Mathews (who was to take a top job in Gemini), on February 6. By now, the Space Task Group people were firmly behind LOR, but von Braun's group still preferred Earth orbit rendezvous.

The early months of 1962 saw continued studies to determine the best method of achieving a manned lunar landing within the end-of-decade deadline set by Kennedy. MSFC continued its studies of EOR versus direct ascent, and MSC concentrated on LOR. NASA also hired outside contractors such as Space Technology Laboratories (a subsidiary of Thompson Ramo Wooldridge Inc.) to study direct ascent. In January 1962, JPL issued Part 2 of its original TM No. 33-52, *An All-Solid-Propellant Nova Injection Vehicle System for the Direct-Ascent Man-on-Moon Project*. This studied the launch vehicle design (solid rocket motor casings, nozzle configuration and material, etc), its performance, its means of production including propellant processing and loading into the casings, the transportation of the stage components to Cape Canaveral, the assembly of the vehicle, and the launch facilities that were to be built 6 to 12 miles offshore on massive platforms. (The Launch Operations Directorate was seriously looking at this possibility, along with land-based launching from Cape Canaveral.) This Nova vehicle would have seven solid rocket motors in its first stage, and would be capable of delivering a 59 metric ton spacecraft to the Moon. The immense size and weight notwithstanding, JPL deemed the building and launch of such a vehicle to be technically feasible.

On February 14, 1962 NASA signed a contract with Boeing for "indoctrination, familiarization and planning, expected to lead to a follow-on contract for design, development, manufacture, test, and launch operations" of the S-IC first stage of the Saturn C-5. Four weeks later, NASA Administrator James E. Webb sent a letter to President Kennedy and Vice President Johnson requesting that the Apollo program be given the highest national priority. On April 11, the President approved National Security Action Memorandum No. 144 granting DX priority to the Apollo program, the Saturn C-1 and Saturn C-5, together with all necessary support to achieve the new national goal of landing men on the Moon and returning them safely to Earth. One week later, NASA announced this news. In mid-April, MSC representatives gave a detailed presentation to MSFC on the benefits of LOR. On April 24, Milton W. Rosen, MSC's Director of Launch Vehicles and Propulsion, recommended the Saturn C-5 be specifically designed to achieve a manned lunar landing using LOR, with the S-IVB designed as the third stage specifically to support this mission mode. The following day, April 25, SA-2 was successfully launched from Cape Canaveral.

On June 7, 1962, a pivotal presentation took place at NASA Headquarters before Dr. Joseph F. Shea, regarding the mode selection for the lunar landing program. The meeting took most of the day, and as it drew to an end Dr. von Braun presented his recommendation and concluding remarks:

> In the previous six hours we presented to you the results of some of the many studies we at Marshall have prepared in connection with the Manned Lunar Landing Project. The purpose of all these studies was to identify potential technical problem areas, and to make sound and realistic scheduling estimates. All studies were aimed at assisting you in your final recommendation with respect to the mode to be chosen for the Manned Lunar Landing Project.
>
> Our general conclusion is that all four modes investigated are technically feasible and could be implemented with enough time and money. We have, however, arrived at a definite list of preferences in the following order:
>
> 1. Lunar Orbit Rendezvous Mode—with the strong recommendation (to make up for the limited growth potential of this mode) to initiate, simultaneously, the development of an unmanned, fully automatic, one-way C-5 logistics vehicle.
> 2. Earth Orbit Rendezvous Mode (Tanking Mode).
> 3. C-5 Direct Mode with minimum size Command Module and High Energy Return.
> 4. Nova or C-8 Mode.

Von Braun stated emphatically that a mode decision had to be made by NASA no later than July 1 if the program schedules, contracts, and remaining spacecraft and

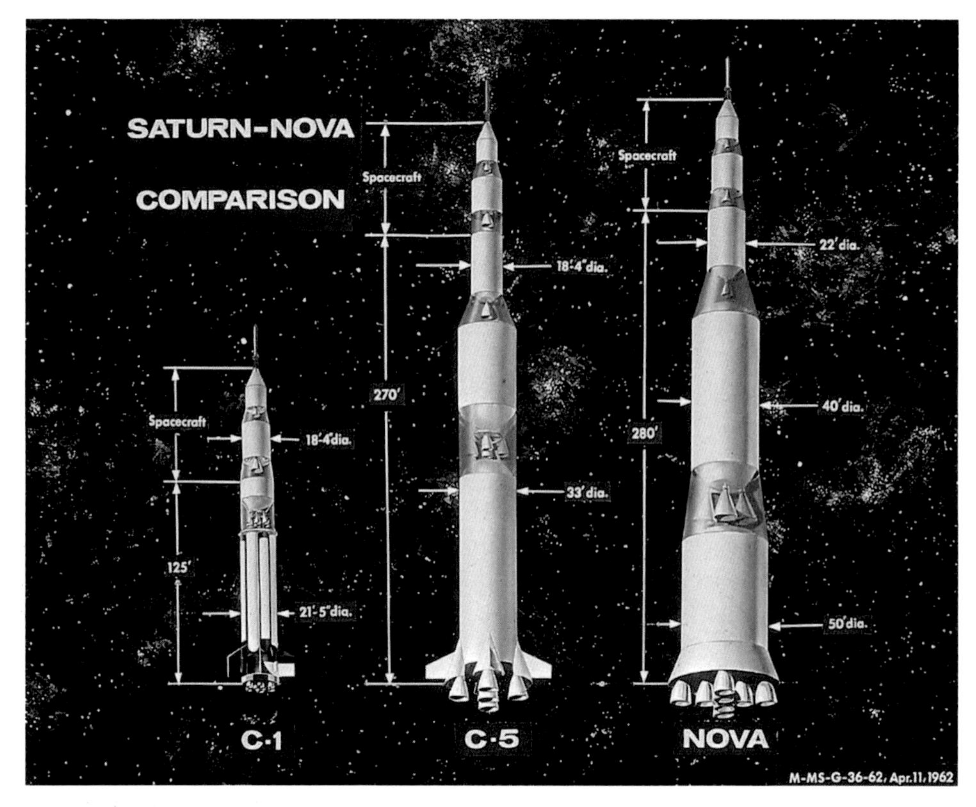


1-11 This illustration from April 1962 shows the comparative sizes of the Saturn C-1, C-5 and Nova launch vehicles. (NASA/MSFC)

launch vehicle assembly and launch infrastructure were to be completed in time to meet the proposed Apollo goal of a lunar landing before the end of the decade. Von Braun gave the compelling reasons for LOR: (1) he felt this mode offered the "highest confidence factor of accomplishment in this decade," and (2) that it offered adequate performance margin.

> We understand that the Manned Spacecraft Center was also quite skeptical at first when John Houbolt of Langley advanced the proposal of the Lunar Orbit Rendezvous Mode, and that it took them quite a while to substantiate the feasibility of the method and finally endorse it.
>
> Against this background it can, therefore, be concluded that the issue of "invented here" versus "not invented here" does not apply to either the Manned Spacecraft Center or the Marshall Space Flight Center, that both Centers have actually embraced a scheme suggested by a third source. Undoubtedly, personnel of MSC and MSFC have by now conducted more detailed studies on all aspects of the four modes than any other group. Moreover, it is these two Centers to which the Office of Manned Space Flight would ultimately have to look to "deliver the

goods." I consider it fortunate indeed for the Manned Lunar Landing Program that both Centers, after much soul searching, have come to identical conclusions. This should give the Office of Manned Space Flight some additional assurance that our recommendations should not be too far from the truth.

Two weeks later, the Manned Space Flight Management Council unanimously decided on lunar orbit rendezvous using the Saturn C-5 as the launch vehicle. But it also recommended that studies of the Nova launch vehicle should continue in anticipation of this rocket being developed after the Saturn C-5 became operational.

## SATURN DIRECT ASCENT STUDIES

The scales were tipping away from direct ascent towards lunar orbit rendezvous, but the feasibility and advantages of direct ascent were still being evaluated. On June 25 1962 Space Technology Laboratories delivered its *Direct Flight Schedule and Feasibility Study for Project Apollo*. In addition to studying "the schedule aspects of the direct flight mission when the Saturn C-5, Saturn C-8 and Nova are considered as alternate launch vehicles", the company addressed "The definition of a spacecraft and its elements suitable for the C-5 direct flight mission." At over 300 pages, this report thoroughly examined all aspects of the proposed launch vehicles, spacecraft, and production schedules. The initial findings indicated a Saturn C-5 with hydrogen–oxygen upper stages and a lightweight spacecraft could provide satisfactory payload margins to facilitate a direct flight approach with a single Saturn C-5. There was no specialized lunar landing craft. Instead, the third stage would essentially serve as the lunar landing stage—itself having a deorbit stage possessing three Pratt & Whitney RL-10A-3 hydrogen–oxygen engines for the initial part of the descent. It would be jettisoned at an altitude of 1,000 meters, and the vernier propulsion stage would land the spacecraft. This approach was proposed to permit a lower center of gravity of the landing vehicle. The ascent stage would have a single RL-10A-3 engine that would also power the spacecraft out of lunar orbit and back to Earth. This vehicle design study assumed a translunar mass of 40.8 metric tons. The study also looked at Saturn C-8 and Nova configurations. The S-IC stage of the C-8 would be powered by eight F-1 engines, the S-II by five J-2 engines (with this stage of a C-8-Prime having *nine* J-2 engines) and the S-IVB having a single J-2 engine. Although the first stage of the Nova would be larger, it would still be powered by eight F-1 engines. However, its second stage would have two Aerojet M-1 engines. The third stage would have a single J-2 engine. The C-8-Prime and Nova were calculated to have a liftoff mass of 4,082 metric tons. The translunar payload was projected at 58.65 metric tons for the C-8, 64.7 metric tons for the C-8-Prime, and 77.56 metric tons for the Nova. The study determined that there would be only a four month difference in delivery of the C-5 and C-8, with a manned lunar landing being feasible during the summer of 1967 using the C-5. The report stated that the Nova would push the first lunar mission out 20 to 24 months beyond that of the C-5.

Space Technology Laboratories issued a follow-up report in February 1963 that examined a reduced diameter command module for two astronauts, as well as the previous three man command module. The two-man mission would last eight days, while the three-man one was ten days in length. As before, the lunar landing stage was made up of the deorbit stage, the lunar landing stage, the lunar liftoff stage and the command module. However, by the time that this report was issued NASA had already decided on the mode for the manned Apollo missions to the Moon.

## LUNAR ORBIT RENDEZVOUS IS CHOSEN

At a press conference at NASA Headquarters on July 11, 1962, Administrator James E. Webb announced that lunar orbit rendezvous would be the mission mode for the Apollo program. John Houbolt was out of the country at the time, presenting a paper before NATO's Advisory Group for Aerospace Research and Development in Paris. He learned of NASA's decision from his division chief, Dr. Isadore Garrick, who read it in the *International Herald Tribune*. Houbolt could hardly believe it. His long struggle had been vindicated.

As von Braun wrote in *Saturn the Giant*, as chapter 3 of *Apollo Expeditions to the Moon* (NASA SP-350, 1975):

> As all the world knows, the LOR was ultimately selected. But even after its adoption, the number of F-1 engines to be used in the first stage of the Moon rocket remained unresolved for quite a while. H.-H. Koelle, who ran our Project Planning Group at Marshall, had worked out detailed studies of a configuration called Saturn IV (Saturn C-4) with four F-1's, and another called Saturn V (Saturn C-5) with five F-1's in its first stage. Uncertainty about LM weight and about propulsion performance of the still untested F-1 and upper-stage engines, combined with a desire to leave a margin for growth, finally led us to the choice of the Saturn V configuration.

Konrad Dannenberg, a veteran member of the German rocket team, was Deputy Manager of the Saturn V at MSFC. He remembers the issue of the number of F-1 engines that would ultimately be required. As he recalled in an interview with this author in 2006:

> Initially, the Saturn [V] had only four F-1 engines. The Houston people—the Space Task Group—had promised that their lunar landing equipment would weigh below 100,000 pounds [45.4 metric tons]. These things always increase in weight and become heavier than you initially think they will be. When von Braun saw the weight was getting close to 100,000 pounds, he made a last minute decision during the contract negotiations with Boeing to add the fifth engine because he predicted that the lunar lander would weigh quite a bit more than 100,000 pounds. The lunar lander finally came out at about 125,000 pounds.

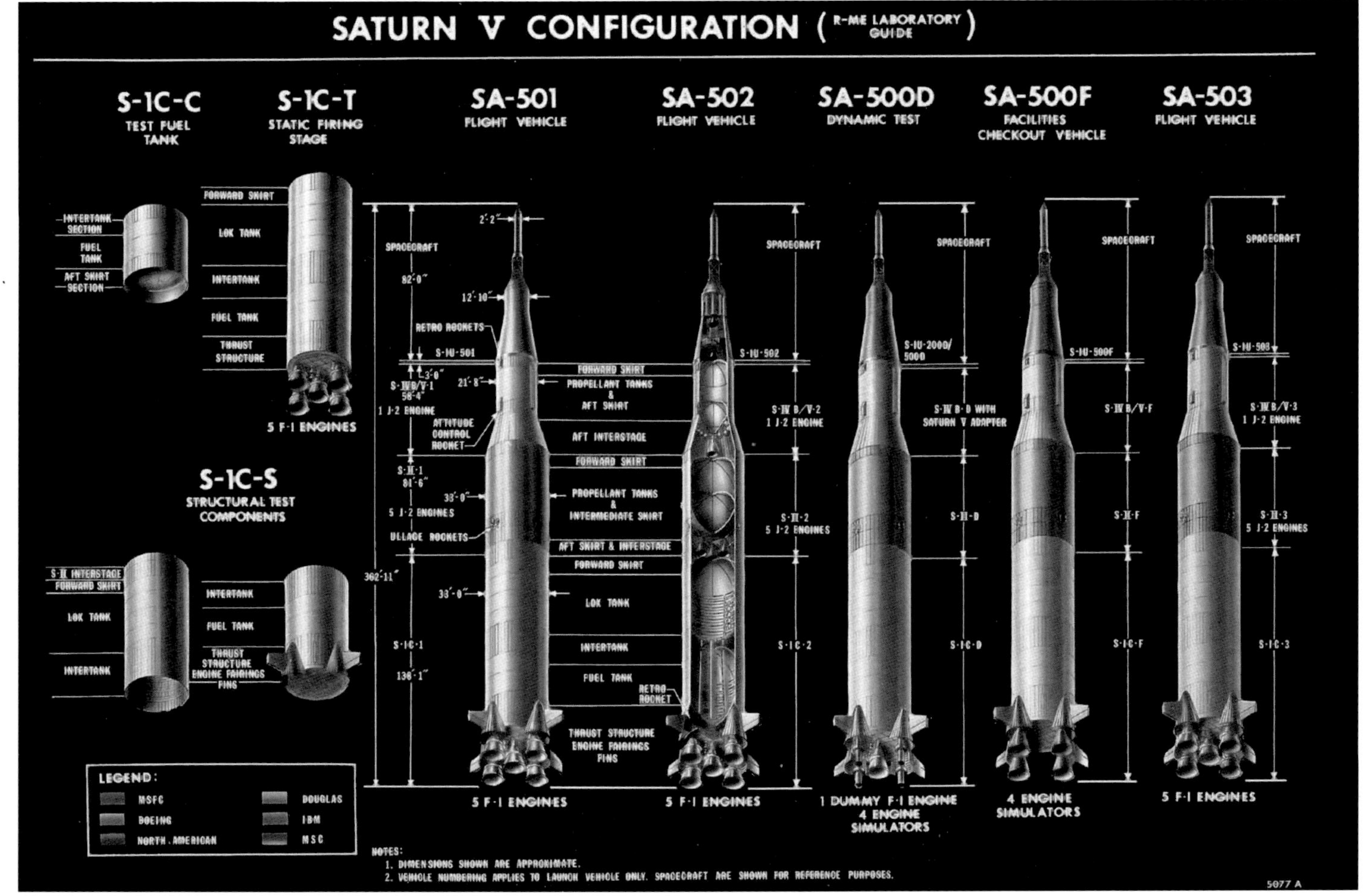


1-12 Several non-flight Saturn V first stages were required, including the S-IC-D Dynamic Test Vehicle, the S-IC-F Facilities Checkout Vehicle, and the S-IC-T Static Firing Test Vehicle. (NASA/MSFC)

Also in June 1962, NASA issued its request for proposals for a static test stand for the S-IC stage to be built at MSFC. The previous month, Rocketdyne had conducted a long duration full-thrust test of the F-1 engine. Much of the Apollo infrastructure was being designed or under construction at NASA facilities around the nation, as well other contractor locations. In September, President Kennedy and Vice President Johnson paid a visit to MSFC. Dr. von Braun was there to greet them as they stepped off Air Force One. Kennedy toured the facilities for two days, and witnessed a test firing of the Saturn I booster. This meeting between the President and von Braun was a defining moment in von Braun's career, and an affirmation of the trust the President and the country placed in his direction of MSFC in its crucial role in the development of the mighty S-IC stage and the management of the F-1 engine program. On February 7, 1963, NASA formally dispensed with the alpha-numeric designations for the Saturn launch vehicles: the Saturn C-1 became the Saturn I, the Saturn C-IB became the Saturn IB and the Saturn C-5 became the Saturn V.

The United States of America was now on an unstoppable trajectory toward the accomplishment of the astonishing goal of landing men on the Moon, investigating the nature of its surface and returning safely to Earth. Apollo would tax the scientific, engineering, manufacturing and management abilities of hundreds of thousands of Americans. Yet doing so would make it an economic and technological superpower. The nation would also be traumatized by the assassination of President Kennedy on November 22, 1963, rocked by country-wide college and university campus protests to the war in Vietnam and enlightened by the women's liberation movement. During this time millions of Americans would mark some of their proudest moments simply by following the news of the Mercury, Gemini and Apollo programs.

To achieve that goal of landing astronauts on the Moon, however, would take the development of the F-1, the most powerful rocket engine the world had yet seen, with a cluster of five of those engines powering the S-IC first stage of the Saturn V.

# 2

# Origin and development of the F-1 engine

What prompted the United States to develop large, single-chamber rocket engines in a process which led to the largest and most powerful liquid propellant engine ever built: the F-1. In trying to sift through the historical events of the late 1940s and the 1950s in rocket engine and missile development to determine the path that led to the F-1, one must focus on the Navaho intercontinental-range cruise missile. At the end of World War II in 1945, the Technical Research Laboratory of North American Aviation (NAA) in Long Beach, California submitted proposals to the Army Air Force to perform research and development on German V-2 rockets. This included first adding wings to the rocket, then the inclusion of turbojet-ramjet propulsion, and finally adding a booster to enable the upper stage payload to achieve intercontinental range. The Air Force issued study contracts for the first iteration of this concept: the MX-770. In April 1947, NAA received the development contract for the SSM-A-2, named Navaho. This program not only had a profound impact on the fortunes of the company, it also advanced the science of missile propulsion, guidance, and aspects of subsonic, trans-sonic and supersonic flight. To conduct this large program, NAA needed new and different facilities from those used for decades in the development of propeller-driven aircraft.

Leland "Lee" Atwood joined NAA in 1934, and became closely involved in the P-51 Mustang fighter and B-25 Mitchell medium bomber. He rapidly rose through the engineering and management ranks, becoming company vice president in 1941 and president in 1948—just as NAA started to seek the appropriate geographic area to conduct the Navaho and potential follow-on programs. As Atwood recalled in an interview conducted in 1989:

> So we began to look for land and a good facility, where a good facility could be erected. Land was much easier to get in those days, around here. In fact, at the end of the war, there was kind a flattening out in all kinds of activity in the commercial sense, so we located this land at Santa Susana in San Fernando Valley. It was

2-1 In engineering and building the Navaho missile for the Air Force, North American Aviation developed rocket engine propulsion systems that laid the groundwork for the engines that powered the Redstone, Jupiter, Atlas and Thor missiles and later the Saturn I and Saturn V launch vehicles. (NASA/KSC)

controlled by the Dundas family. It had been used for nothing more useful than some 'cut them off at the pass' Western movies. It had some defile and rocky escarpments that were pretty good for movie sets, but the land didn't seem worth much. The Dundas family eventually sold us about a square mile up there for

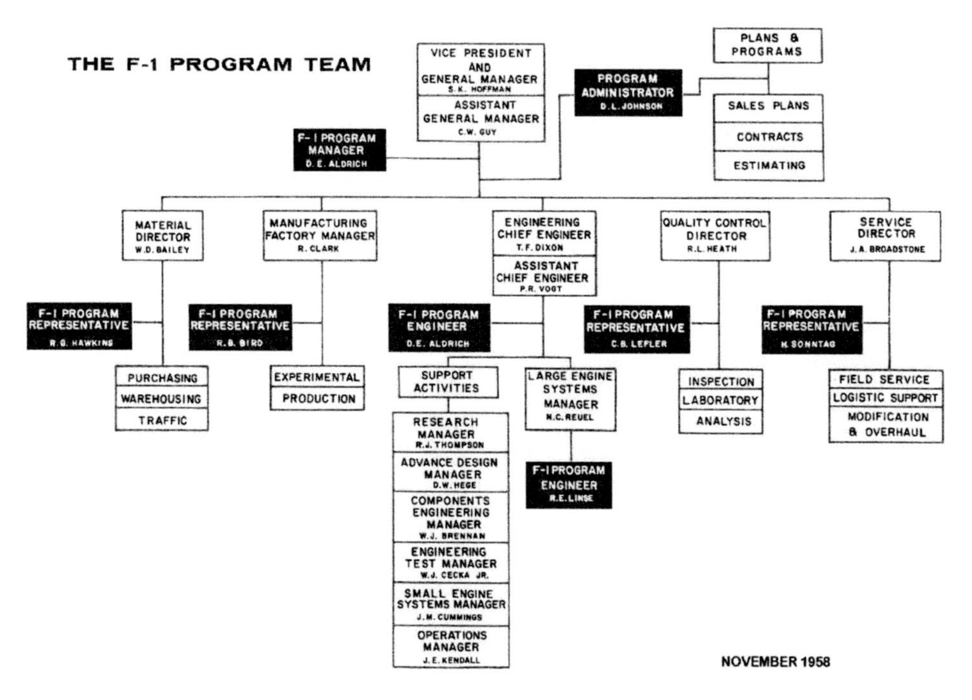


2-2 The F-1 Engine Program Team organized in 1958 drew on some of the most experienced engineers and managers at Rocketdyne. (Rocketdyne, Vince Wheelock Collection)

some $300 an acre, and we began to build a rocket test stand up there. It was pretty well isolated from residential areas and had a kind of bowl effect, and the sound didn't propagate directly into the neighbors' windows or anything like that, so I felt we were pretty lucky to be able to get a test facility that close to general civilization.

The next move in the rocket field was to start a plant and our office organization structure over in Canoga Park. It was about the time they were beginning to break up the old Warner Ranch, and so we were able to get property there, and that's where Rocketdyne's headquarters is today. It was, I would guess, four or five miles from the rocket test facility up a winding road. So that facility went on from there.

NAA was not the only firm winning Air Force contracts for missile studies. The Vultee Field Division of the Consolidated-Vultee Aircraft Corporation of Downey, California secured a contract for a missile program to study both cruise and ballistic modes. Vultee settled on the ballistic mode. Its first proof-of-concept was the MX-774. Several were built and launched at the test range at White Sands, New Mexico. Work on the MX-774 was suspended by the Air Force after two and a half years, but the experience gained was eventually exploited when Convair won the contract for the MX-1593 intercontinental-range ballistic missile (ICBM) named

Atlas. In order to keep its options open, the Air Force pursued both cruise missiles like the Navaho, Snark and Matador, among others, and ballistic missiles such as the Atlas and Thor. Meanwhile, the Army was exploiting its experience with V-2 technology to develop the Jupiter intermediate-range ballistic missile.

Even in the midst of the preliminary work on the Navaho, NAA was involved in other studies for the U.S. Government to provide a launch vehicle capable of placing a satellite into orbit. The design studies, with contracts administered by the RAND Corporation, would eventually lead to development contracts for such a vehicle. In October 1945, the Navy's Bureau of Aeronautics had established the Committee for Evaluating the Feasibility of Space Rocketry. The Bureau of Aeronautics contracted with NAA in 1946 for a 90-day study of the feasibility of the Bureau's proposal for a launch vehicle based on structural mass and propulsion figures which had been provided by JPL/GALCIT, which itself had done a study for the Bureau. The Navy also requested a study from the Glenn L. Martin Company. The NAA study was for a launch vehicle with a hydrogen-oxygen engine that could place a 454 kg satellite into orbit with a single stage with a total initial mass of 59,000 kg. NAA planned to achieve this by using pressure-stabilized propellant tanks, with a common bulkhead between the liquid hydrogen and oxygen tanks. Although this study did not result in a launch vehicle development contract from the Navy, the company was able to use the development of the Navaho as its technology driver.

The Navaho had a liquid propellant booster stage with the ramjet cruise missile mounted on its back, in shuttle-like fashion. To develop and test the rocket engine, NAA constructed the first Vertical Test Stand (VTS-1) at the new Propulsion Field Laboratory in the Simi Hills, later name the Santa Susana Field Laboratory, in the area that became known as The Bowl. The power plant designed for this missile was a dual-engine propelled by liquid oxygen and alcohol. When fully developed, each engine was capable of delivering 120,000 pounds of thrust and had a specific impulse of 65 seconds. This engine had the designation XLR71-NA-1. The complex Navaho program went on for years, and suffered a variety of hardware performance problems. The first development flight in November 1956 ended in failure less than a minute after launch from Cape Canaveral. Three other development flights were made the following year, before the program was cancelled in July 1957. The future of missile technology was clear from the facilities constructed nearby, most notably Launch Complex 14 for the Atlas ICBM, the first flight of which was made on June 11, 1957. Nevertheless, the Navaho program was instrumental in the development of rocket propulsion technology that would later lead to the mighty F-1 engine.

As Atwood observed in his interview:

> The Navaho added capability. It added some profit, it stretched organizational capabilities—it created, really, the rocket engine business. It created the guidance system business which we still have, in considerable strength. It augmented our aerodynamic information on high speed aerodynamics and control. The wind tunnel testing done and the actual testing are the things that improved our structural capabilities in lightweight structures, that sort of thing.

2-3 Samuel K. Hoffman was Vice President of North American Aviation from 1957 and General Manager of Rocketdyne since 1955 during the F-1 engine program. (Rocketdyne, Vince Wheelock Collection)

## THE E-1 ENGINE

In tracing the Rocketdyne F-1 family tree, one discovers something of a mystery in the form of the E-1 engine. The early Saturn I configurations conceived by Heinz-Hermann Koelle, Director of Future Projects at MSFC, called for four E-1 engines, each generating 400,000 pounds of thrust, to provide the required 1.5 million pounds of thrust at liftoff with a comfortable margin. However, the E-1 was never utilized in either the Saturn I or any other launch vehicle. Nevertheless, hot fire tests of the E-1 were conducted, and photographs of these tests prove its existence. The brief life of the E-1 and its influence on the F-1 is very interesting.

In July 1954, prior to the establishment of Rocketdyne as a separate division of NAA, the ICBM Scientific Advisory Committee (SAC), which comprised leading aerospace industry engineers and civilian scientists, recommended to the Air Force's Western Development Division (WDD) that a missile with a configuration different to the Atlas, which was then under development, should be considered as a backup option. This was because the ICBM SAC was skeptical of the Atlas's airframe and unconventional propulsion system. The WDD duly directed the Ramo-Wooldridge Corporation, which was its systems engineering and technical direction contractor, to make a study of alternative ICBM designs. The company invited the Lockheed Aircraft Corporation and the Glenn L. Martin Aircraft Company to offer ideas for an alternative design. The results were submitted to WDD for evaluation. The WDD agreed that a second ICBM program should be initiated, employing a rigid airframe and a different booster engine configuration. The ICBM SAC duly recommended to the Air Force that it establish a parallel ICBM program for a two stage missile with a rigid airframe. This program was approved by the Air Force, and contracts awarded to the Glenn L. Martin Aircraft Company for the airframe and to

2-4 David E. Aldrich was Rocketdyne Program Manager and Program Engineer of the E-1 and F-1 engine programs at the time that the company developed its *One Million Pound Liquid Propellant Rocket Engine Proposal.* (Rocketdyne, Vince Wheelock Collection)

Aerojet General for the propulsion system burning RP-1 and LOX. The new missile was given the name Titan.

The twin-chamber LR-87-3 engine supplied by Aerojet General for the first stage of the Titan produced an overall thrust of 327,800 pounds (1,447 kN). As part of the WWD's conservative approach to the ICBM program, the new Rocketdyne division of NAA, which was established in 1955, was chosen as the backup contractor for the Titan propulsion system. Rocketdyne's Advance Design Group opted to depart from the multi-engine configurations of the Navaho and Atlas. On receiving the contract to proceed with development, the company decided to make a large, single-chamber regeneratively cooled engine burning RP-1 and LOX with a target thrust of between 300,000 and 400,000 pounds. This was designated the E-1. It resembled the S-3D of the Jupiter, but was much larger and the thrust chamber had a more pronounced curvature. Development testing at the Santa Susana Field Laboratory over a period of months resulted in a maximum thrust in excess of 379,837 pounds (1,689 kN) at sea level. The first mainstage test occurred on January 10, 1956. Impressed with the E-1, Heinz-Hermann Koelle proposed that it be used for the Saturn I, but this would not be realized. In explaining to this author in 2007 the sequence of events relating to the E-1 and the Saturn I, Koelle wrote:

> Being aware of the booster gap, I started the design of the JUNO V in April 1957, six months before Sputnik I was launched. I was shooting for a one million pound launch mass and a 10,000 kg payload capability to low Earth orbit. For this I needed a propulsion system. In my frequent contacts with Rocketdyne, particularly with the Preliminary Design Section headed by George Sutton, I learned about their E-1 engine project, which at that time was the largest engine under consideration for the near future. Thus, I selected a four engine E-1 cluster for the first stage of the JUNO V. In July of 1958, two ARPA engineers, Dick Canright and Bob Young, came to see Dr. von Braun, telling him that they had $10 million left before they had to turn over all launcher development to NASA in November. They asked him whether he could use this money for advanced booster development. He called me in. I presented an impressive detailed booster drawing, scale 1:10, of the JUNO V. Dr. von Braun pointed out that our new test stand could handle this booster with a few reinforcements. Canright and Young proposed to take eight Jupiter engines instead of the four E-1 engines, because these engines could be available much sooner and would be cheaper. We agreed to this, and a month later we received the authority to proceed with the development of a demonstration booster of the proposed size. In December we were able to convince the NASA leadership by a presentation in Washington that we had a sound program, and they agreed to continue financing this development.

Aerojet successfully established the LR-87-3's reliability, and the first production Titan missile was delivered to the Air Force in 1958. Rocketdyne requested from the Air Force permission to terminate the E-1 development, and this was approved. The engine was never in production. The Future Projects Office and the Engine Program Office at MSFC gave some thought to taking over the development of this engine but,

2-5 Robert E. Linse was F-1 Project Engineer during the early years. (Rocketdyne, Vince Wheelock Collection)

as von Braun admitted years later, the cost would have been prohibitive. In any case, events would soon overshadow the E-1. The Juno V became the Saturn I with the more economical H-1 engine, and attention switched to a far more powerful and much larger engine known as the F-1.

## EARLY F-1 ENGINE STUDIES

In the early to mid-1950s, Rocketdyne's Advanced Design Group investigated the theoretical limits of rocket engines, in terms of their size and thrust. To place a very heavy payload into orbit, or send men to the Moon, would need much larger engines than existed at that time. As related in the Introduction, during this period both the Army and the Air Force contemplated large payloads, including manned spacecraft, which would require an engine delivering as much as 1 million pounds of thrust. In 1955, NAA formed four divisions: Missile Development (later Space Systems), Autonetics, Rocketdyne and Atomics International. Rocketdyne won the contract for the booster and sustainer engines of the Atlas ICBM for the Air Force, and also the engine for the Army's Redstone medium-range ballistic missile. In December 1955 it gained the contract to produce the engine for the Thor intermediate-range ballistic missile that was being made by the Douglas Aircraft Company for the Air Force. As a result of these and other programs, and the perceived need for ever larger engines, Rocketdyne grew rapidly. In November 1956 the fuels and propulsion panel of the Air Force's Scientific Advisory Board investigated the likely requirements for large liquid propellant rocket engines, and recommended a study of engines delivering up to 5 million pounds (22.3 MN) of thrust! The Air Force took its time considering the

2-6 Paul Castenholz was placed in charge of the Rocketdyne team that resolved the combustion instability problems with the F-1. (Rocketdyne, Vince Wheelock Collection)

recommendation, and doubtless invited inputs from Rocketdyne and the other rocket engine companies.

The first question was whether such a massive engine could be built and operated reliably. Drawing on their experience, ranging from the Navaho through to the E-1, the Rocketdyne engineers applied analytical methods to all aspects of this theoretical engine design, and in 1957 the company reported to the Air Force that there were no insurmountable difficulties. Although the Air Force had no immediate requirement for such an engine, this forward-thinking branch of the military felt there might be a need for an engine with 1 million pounds of thrust in the near future. The launch of Sputnik 1 by the Soviet Union on October 4, 1957, and the much larger and heavier Sputnik 2 a month later, put the U.S. on notice that its ideological enemy had vastly superior boosters. In June 1958 the Air Force contracted Rocketdyne to undertake the preliminary development of the hardware for such an engine, making subsequent funding contingent on successful demonstration tests.

The Program Management Team formed by Rocketdyne to work on this engine was structured along a line organization that had proved itself with other programs. David E. Aldrich of the Rocket Engine Advancement Program (REAP) had served as both Program Manager and Program Engineer on the E-1 engine, and was named Program Manager and Program Engineer for the F-1. He would report directly to Samuel K. Hoffman, who was not only General Manager of Rocketdyne but also Vice President of NAA. Robert E. Linse, who had been Project Engineer on the E-1 and other Rocketdyne programs, was named Project Engineer for the F-1. Douglas W. Hege joined the company in 1947, served as Project Engineer for the Navaho, Redstone, Jupiter, Atlas and Thor, then Program Manager for the Atlas propulsion system. He was currently Manager of Advance Design and also involved with the F-1 program. A key member of the team, Dr. Robert J. Thompson, was Manager of

Research for all Rocketdyne programs. William J. Brennan, Program Manager of Components Engineering, welcomed the challenge of managing the engineering of the largest liquid propellant rocket engine components ever conceived. Norman C. Reuel, who joined the company in 1946, was Manager of Large Engine Systems. William J. Cecka, who had also been with the company 12 years and was Manager of Test Engineering of the Propulsion Field Laboratory, was responsible for testing the F-1 components, and eventually the engine itself. Ted Benham was added to the team during the development phase as Project Engineer. Despite the size of the F-1 engine, Rocketdyne's approach to its design emphasized simplicity, ruggedness and reliability. That simplicity began with the method of starting the F-1.

As Ted Benham recalled to this author in 2007:

> When we first got into figuring out how we were going to develop the engine, we first had to figure out how we were going to start it. There was a little experimental work going on that was called the tank head start. That's where you utilize the tank pressures to start the engine. I made the decision that that's the way we were going to go. That was a big step, because we didn't have much information or history on that type of thing—because it was an experimental idea.

The F-1 was conceived by Rocketdyne to use the gas generator cycle, whereby propellants are burned in a gas generator to drive the turbine of the turbopump. This had been done on the Redstone, Thor and Jupiter engines, and would also be used on the H-1 and J-2 engines. As a further step to simplify the F-1, Rocketdyne did away with individual gear-driven fuel and oxidizer pumps, and instead placed both pumps on the shaft of the turbopump in order that the oxidizer pump, fuel pump and turbine shared a common drive. The exhaust gas from the turbine would pass through a heat exchanger containing coils holding LOX and helium, which, by expanding, would pressurize the propellant tanks. The heat exchanger's exhaust would by discharged into the extension of the thrust chamber to provide a cooler boundary layer to protect the wall from the hot exhaust from the thrust chamber. On June 23, 1958 the Air Force hired Rocketdyne to develop a single-chamber engine that would burn RP-1 and LOX and deliver 1 to 1.5 million pounds of thrust. A month later, it awarded the development contract. However, events in Washington, D.C. would soon change the course and pace of the F-1 engine.

## TRANSFER OF THE F-1 ENGINE PROGRAM TO NASA

On July 29, 1958 President Dwight D. Eisenhower signed the National Aeronautics and Space Act, which legislatively established the National Aeronautics and Space Administration. On the recommendation of his advisors, Eisenhower nominated T. Keith Glennan as Administrator and Hugh L. Dryden, head of the National Advisory Committee for Aeronautics, which was to be integrated into NASA when this became operational on October 1, as Deputy Administrator. Even as NASA's management was being formed, Glennan and Dryden knew that rocket propulsion

2-7 Ted Benham in front of an early mockup of the F-1. (Rocketdyne, Vince Wheelock Collection)

was one of the key drivers in gaining access to space, both for unmanned and manned and missions. Among the organizers of America's new space agency was Dr. Abe Silverstein, who was head of the Lewis Aeronautical Laboratory. He was well aware of the need for larger rocket engines, and the budget he submitted on July 19 for the future agency included $30 million to initiate development of a 4.5 MN single-chamber engine, as well as $15 million for development work on clustering currently available ICBM engines in order to attain the same level of thrust. Silverstein organized a propulsion committee in August, and at a meeting that month in Washington the Air Force fully briefed Silverstein on its contract with Rocketdyne for the 4.5 MN engine, pointing out that by September or October the F-1 program would exhaust its development funds. Additional funding would be needed to continue the work, and it was evident to Silverstein that NASA would have to make a significant contribution. At another meeting on August 28 the Air Force informed Silverstein that contracts would be let for a test stand at Rocketdyne's Santa Susana Field Laboratory suitable for a 4.5 MN engine. There was a test stand at Edwards Rocket Test Center capable of handling that much thrust, but it was being used by the Atlas program, which had the highest national priority. Adelbert O. Tischler, who advised Silverstein on propulsion, was present at this meeting. Although Rocketdyne had invested development time in the F-1, Silverstein and Tischler did not wish to limit NASA to a sole-source contractor, since as part of its basis the agency was to utilize competitive bidding to evaluate different approaches to design, development, testing and manufacturing. Silverstein announced at the meeting that any development of a large rocket engine like the F-1 would be initiated by competitive bidding. This did not worry Rocketdyne's General Manager, Samuel K. Hoffman, as he was confident that his company would win the bid. On September 11, 1958 a letter contract was signed by NASA and Rocketdyne to develop the smaller H-1 engine that was to be clustered to produce 1.5 million pounds of combined thrust for the booster stage of the proposed Juno V (later named the Saturn I).

When the Silverstein Committee met on October 9, 1958 Tischler announced that within one to two weeks NASA would open competitive bidding for a large, single-chamber engine rated at 1.0 or 1.5 million pounds of thrust (4.7 or 6.7 MN). In due course, the agency sent invitations to bid to seven contractors. At a bidders briefing in Washington D.C. one week later, Tischler said the designated target thrust was to be 6.7 MN. On November 10, Rocketdyne submitted to NASA its *Program for the Development of a 1,000,000-Lb-Thrust Liquid-Propellant Rocket Engine System.* In fact, this contained eight individual reports:

- Summary (R-1170P)
- Development Program (R-1171P)
- Experience (R-1172P)
- Facilities (R-1173P)
- Studies and Research (R-1174P)
- Manufacturing Techniques (R-1175P)
- Management Plan (R-1176P)
- Model Specification (R-1177S).

On November 24, NASA set up a technical and a management team to evaluate the contractor proposals, and on December 9 these teams submitted their reports to the Source Selection Board, which reviewed the recommendations and in turn sent a recommendation to the Administrator that the development contract for this engine be awarded to Rocketdyne.

On December 12, 1958 NASA announced the selection of Rocketdyne to develop the F-1 engine under a contract that was in the process of negotiation. On January 19, 1959, it signed a definitive contract with the company to design and develop a single-chamber RP-1 and LOX engine capable of developing 1.5 million pounds of thrust. The contract was for $102 million. At that time, the idea was that the engine would be clustered for the proposed Nova launch vehicle for a direct ascent mission to the Moon.

## THE ROAR OF 'KING KONG'

Among the first pieces of hardware for the F-1 engine to begin development was the injector. Although Rocketdyne naturally drew on its experience of previous engines burning RP-1 and LOX, the size of the F-1 injector was unprecedented.

2-8 Bob Biggs with his model depicting the characteristics of the Jupiter Block III engine. He celebrated his 50th year with Rocketdyne in 2007. (Rocketdyne, Vince Wheelock Collection)

Dan Brevik joined Rocketdyne in August 1956, and told this author in 2006 that since this was a proof-of-concept development program the thrust chamber for the first tests was, in Rocketdyne's parlance, the essence of simplicity:

> In the beginning, we didn't have a thrust chamber built up, we had a mockup. The thrust chamber on the first mockup was solid; typical for rocket development back then—it wasn't even a bell-shape, it was a cone because that was easier to machine. This was done strictly to test the injector dome assembly because that was always the biggest problem to get those things to work.

The first static firing on the high pressure horizontal test stand Bravo 1 at the Santa Susana Field Laboratory was conducted on March 6, 1959. The firing lasted only a few seconds, but the thunderous noise was unlike anything heard from Santa Susana. The test engineers dubbed it 'King Kong'. It rattled windows for miles down

2-9 A mainstage test of the 400,000 pound thrust E-1 at the Santa Susana Field Laboratory in 1959. (Rocketdyne, Vince Wheelock Collection)

the San Fernando Valley and led to telephone calls to Rocketdyne. Even on its first test, the engine had generated in excess of 1 million pounds of thrust. It was an historic moment for Rocketdyne and the F-1 program, and a milestone in the development of rocket propulsion. The test also made apparent that firings of the

F-1 could not continue for long at Santa Susana. Although F-1 component testing would continue at the Santa Susana Field Laboratory, the future home of the F-1 would be Edwards. (See Chapter 7)

## ESTABLISHING THE ENGINE START SEQUENCE

One of the key engineers involved with the F-1 engine program was Bob Biggs. He joined Rocketdyne in 1957 and started work on the Navaho, but this was cancelled three months later. After working on the Jupiter for several years, he was transferred to the F-1 program, where he worked for Stuart Mulliken, Manager of the Engine Systems Development Group. Biggs and Mulliken were responsible for performance analysis, test planning and other aspects of the engine within the group. As Biggs told this author in 2007:

> The idea for engine start was to break away the pump just using LOX flow by opening the main LOX valves—there were two. The LOX flow going through the impeller acts as a turbine and spins up the pump. That provides it with the initial start. With the pump building up pressure as the gas generator builds up, other events happen as it gets to a certain pressure level. At this point it is in closed loop with the LOX valves open, and it is increasing in power gradually—not nearly as fast as it would with the valves closed. As the fuel pressure rose to 300 psi, it would rupture a hypergol cartridge containing triethylaluminum and triethylboron, which would flow into the injector and ignite the LOX already in the thrust chamber. This hypergolic fluid entered into the chamber through ports in the face of the main thrust chamber injector to give uniform and reliable ignition. As the ignition pressure increased to a certain value, the ignition monitor valve would allow hydraulic fluid to go into the main fuel valve, which would then open, on the way to mainstage. The propellant valves were hydraulically actuated with engine fuel in the start using a ground pumping unit. This source was checked off when the engine fuel pressure took over. A small amount of fuel passed through the oxidizer valve actuators to prevent freezing. The ground pump unit also provided for making sequencing checks, and also thrust chamber movement for guidance checks. NASA liked this feature.

## THRUST CHAMBER TUBE BUNDLE DEVELOPMENT

Initial hot fire component development tests of the F-1 employed a short, solid wall combustion chamber. This permitted the preliminary development of the propellant injector dome and plate assembly, turbopump and related hardware. Firing durations were short enough to allow the use of solid wall combustion chambers, which was standard practice at Rocketdyne prior to the design development of the regenerative cooling tube bundles to make up the combustion chamber.

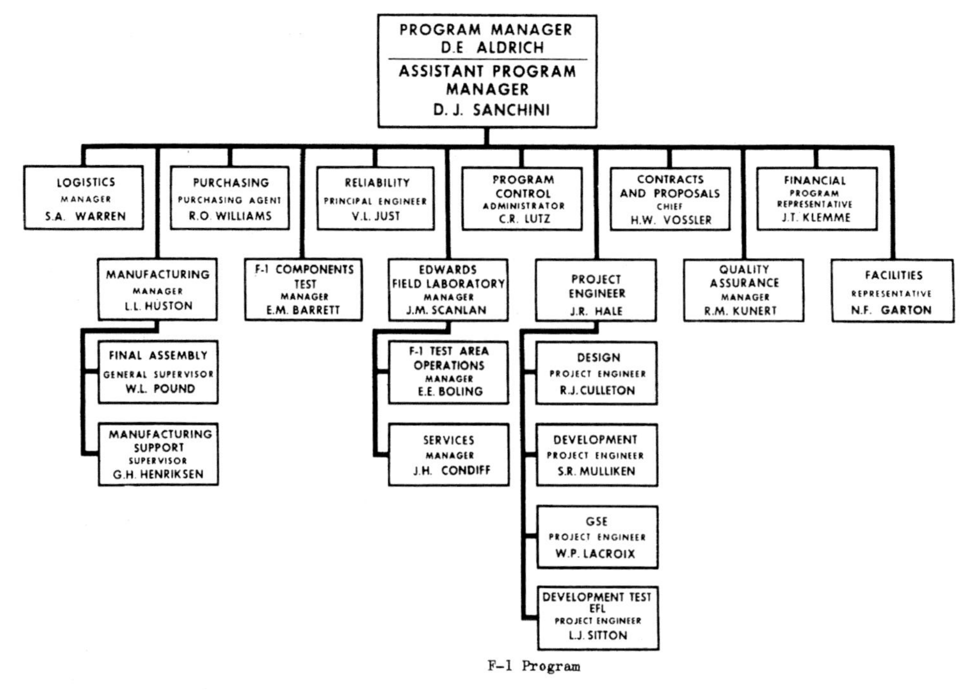


2-10 The organization chart of the F-1 engine program at Rocketdyne in 1961. (Rocketdyne, Vince Wheelock Collection)

After having served as Development Engineer at the Bravo-1B test stand at the Santa Susana Field Laboratory, Dan Brevik was transferred back to the Combustion Devices group in Canoga Park in April 1960. This came about as a result of a report he had written for an assignment given in a rocketry and technology class offered at the University of California at Los Angeles. The assignment was to design a liquid propellant rocket capable of boosting a 1,000 pound payload from low Earth orbit to escape velocity. Brevik submitted a 56-page report with calculations and formulas throughout. A series of events drew Brevik's report to the attention of engineers at Rocketdyne, who were so impressed that they arranged for him to come down from The Hill (as the Field Laboratory was informally known) to join the group that was developing the F-1 engine. On arriving Lou Bononi introduced Brevik to the new engine:

> Lou took me down past the production floor that included Saturn I H-1 engines to see a mockup of the F-1. Now I was used to rockets—I had crawled into enough of them for a lifetime—and we walked past more of them. We were approaching a huge scaffolding draped in canvas, like a large tent. We walked up to it and Lou parted the canvas to let me in. And there was the F-1 mockup in all its glory. I stopped dead in my tracks. The size! Lou said my reaction was about the same as everyone else's.

Brevik was given the task of ensuring the preliminary combustion chamber tube bundle and manifold design produced by Al Bokstellar would run cool. In essence, his orders were to "make sure it doesn't melt." As Brevik recalled:

> Al Bokstellar designed the tubes and bundles for the F-1, and Lou Bononi worked on developing the means of brazing them all together. I started out by doing my calculations on a slide rule and a Monroe Multimatic calculator. That's what we used around then. Then I graduated to the computer and found out what you could do with it. Since no F-1 tube bundles actually existed yet, I learned to program the new computer, which at that time was the most powerful in the world. Rocketdyne got one because our work was so critical to national defense. Using that computer, I calculated the hydrodynamic and thermodynamic characteristics of the F-1; that is, how fluid flowed through it and what the heat transfer was like. I wrote the programs in Fortran. These programs became more and more elaborate. My calculations showed we were looking at a disaster! The tubes themselves were fine. Al had done a good job, but the manifolds that fed fuel into the tubes were all wrong.
>
> I was able to demonstrate heat transfer problems that we were going to get into—a phenomenon called 'starvation.' It occurs where you have the [fuel] manifolds. That manifold feeds the down tubes in the tube bundle. There is a small manifold at the end of the tube bundle where the fuel turns around and goes into the up tube, then into another manifold which feeds into the injector. Bernoulli's Principle says that the faster the fluid flows, the lower the static pressure gets. In the manifold, if the cross-sectional area is too small you will get a very high flow rate at the inlet. As the fuel feeds off into the tubes, the flow rate will go down and the static pressure will go up. What happens then, is you have an imbalance of static pressure which is what drives the fuel down the tubes, and you will get the phenomenon known as starvation. At the feed point, the velocity is so high that the static pressure gets low and it can't push as much fluid through as it does toward the end of the manifold. That means you are going to have hot spots along those tubes since you don't have quite the flow you need.
>
> The thrust chamber tubes were double-tapered with a narrow spot in the middle of the tube and flared out on the ends. The narrow spot was the throat. The round tube was then put in a die so that it was pressed on both sides, squashing it; so the tube then had a cross section of two semi-circles and two flat spots connecting them [i.e. the shape of a slot]. What we were striving for there was a certain cross-section for flow purposes to get the best heat transfer characteristics, and those were calculations done by Boxstellar. The tube material was Inconel-X, which was brand new at that time. It had tremendous properties, but no one had used it before. Lou Bononi had to figure out how to braze that stuff.

Regenerative cooling of the F-1 thrust chamber was indeed a challenge, requiring a careful balance of the amount of RP-1 fuel sent to flow through the thrust chamber tubes with that which was sent to the injector. To verify the calculations and initial

2-11 An F-1 injector test using water. (Rocketdyne, Harold C. Hall Collection)

design of the thrust chamber tubes, tests were conducted using water in one of the test labs at the Canoga Park plant. Brevik recalled this procedure to the author:

> On one occasion, we got a partial F-1 built that had a three to one chamber from the injector face. Before they put on the balance of the tubes, we put it on a water bench and flowed water through it. Those present included Lou Bononi, Don Weldon and I. We determined the flow of the water using a tape to measure how far the water went spurting out the ends of the tubes, because they had no manifolds on them. From that we were able to calculate the flow rates through each down tube and compare those numbers against my program—and it was dead-on.

The material finally selected for the F-1 thrust chamber tube bundle, reinforcing bands and manifolds was Inconel-X750, a refractory nickel based alloy capable of withstanding high temperatures. The thrust chamber was comprised of 178 primary tubes and 356 secondary tubes which were bifurcated with the primary tubes. There was in excess of 3,000 feet of brazed jointing between the tubes. Manufacturing the thrust chamber propellant tubes was a particular challenge, as a very small change in dimension would add up and cause the tubes not to fit the assembly. Therefore, great care had to be exercised in making any tube changes. Considerable effort went into development of the tube die that would expand the tube under very high pressure.

## THE INJECTOR AND COMBUSTION INSTABILITY

The most anticipated factor impacting the stable and reliable performance of a liquid rocket engine is combustion instability. Virtually every engine that Rocketdyne had developed to date had shown varying degrees of combustion instability, and in each case there had been a development program to resolve it. As the instability stemmed from an engine's specific component design and performance, it was rarely possible to apply the solution developed for one engine to another. However, the *knowledge* acquired in resolving one engine's combustion instability contributed to identifying and overcoming instability on later engines.

Rocketdyne prepared in a methodical fashion to deal with combustion instability with the F-1. However, F-1 engine performance goals drove component design, not anticipation of combustion instability. As the E-1 engine was the largest engine that the company had worked on so far, it drew heavily on this experience in tackling the F-1. Design development of the F-1 injector assembly, fuel and oxidizer manifolds, oxidizer dome and combustion chamber injector assembly interface proceeded with the active involvement of the Component Test Laboratories. The first experimental injectors were tested with short, solid-wall combustion chambers at the Santa Susana Field Laboratory. A total of 44 full-scale injector development tests were conducted between January 1959 and May 1960 at thrusts of up to 1,000,000 pounds (4.5 MN). Combustion instability occurred in almost every test, with spontaneous instability in excess of rated chamber pressure in nearly 50 percent of the cases. All injector faces displayed erosion from burning. The engine monitoring equipment on the test stand at Santa Susana included a Rough Combustion Cutoff device, and this intervened with monotonous regularity to prevent the destruction of an engine.

## 'PROJECT FIRST'

The persistent problem of injector combustion instability during 1961 and 1962 was clearly the largest development issue with the F-1 engine, and a looming concern at Rocketdyne and the Marshall Space Flight Center. In October 1962 Rocketdyne's Liquid Propulsion Division established the Combustion Stability Committee, which was charged with the technical direction and management of the F-1 Combustion Stability Program, known as 'Project First'. Paul D. Castenholz was director of both the committee and the program. He was assisted by Dan O. Klute and Bob Fontaine.

2-12 An F-1 injector and thrust chamber tube bundle test using water (Rocketdyne, Harold C. Hall Collection)

The Committee had two primary objectives: (1) to achieve a dynamically stable F-1 engine within the program's schedule and engine specification; and (2) to determine the design and operational parameters that were fundamental to the development of dynamically stable liquid propellant engines. It started with a survey of combustion instability in earlier engines and, in meetings with the F-1 Program Team, undertook a comprehensive examination of development history of the F-1 engine. In order to gain a full understanding of the problem, the Committee decided to investigate every component of the engine that might be a possible source of, or otherwise contribute to, combustion instability; not simply the injector itself. It reviewed the development of the turbopump, propellant lines, injector dome assembly and manifolds, the thrust chamber and the gas generator. And, of course, the evolution of the injectors over the period of the program was closely studied. The combustion instabilities of the Redstone, Atlas, H-1 and E-1 programs were reviewed. The Rocketdyne committee worked with an ad hoc committee assembled by MSFC from industry and academic authorities on liquid rockets and combustion instability (see Chapter 3). Vital to this investigation was the experimental work with a two-dimensional transparent thrust chamber using high speed film analysis. A list of potential sources of combustion instability was compiled, which included:

1. Structural integrity of thrust chamber mounts
2. Thrust chamber nozzle stiffness
3. Structural integrity of baffles in the combustion chamber
4. Characteristics of the material of the thrust chamber
5. Flow straightness in the propellant
6. Diameter of the propellant lines, their length and stiffness
7. Pressure drop in the feed system
8. Velocity changes during operation
9. Cavitation in the injector orifices
10. Cavitation in the feed system
11. Gas pockets in the feed system
12. Diameter of the combustion chamber
13. Length of the combustion chamber
14. Splash plate and turbulence ring disturbances
15. Chamber shape
16. Spacing of injection orifices
17. Temperature of the combustion chamber wall
18. Mixture ratio uniformity.

The Committee examined other areas of concern that included propellant droplet breakup during combustion, baffle theory and acoustic theory.

On the basis of a review of the Combustion Stability Program for the E-1 engine, Volume 1 of the Project First program stated:

> Combustion instability in the E-1 was overcome through the use of wall gap with divergence, and close attention to pattern design that would provide 'adequate'

2-13 A centerline view of an F-1 thrust chamber ready for inspection. (University of Alabama in Huntsville Library Archives)

recirculation at the face. A 50 percent divergence was recommended as a good starting point for adoption of the concept to the F-1. The E-1 experience further indicated that the design parameters affecting performance are chamber length, injector diameter, and pattern.

In December 1962 the F-1 Combustion Stability Group was formed under the direction of the Combustion Stability Committee. It was comprised of an Analysis Unit and a Design Unit.

The Analysis Unit had four areas of investigation: (1) identification of the mode or modes of instability encountered in testing; (2) parametric correlation of test data

2-14 The final injector design to resolve the F-1's combustion instability required this array of baffles. (Rocketdyne, Harold C. Hall Collection)

in order to define stable and unstable modes of operation; (3) analysis of test data on the basis of correlation of theories of instability; and (4) examination and evaluation of exploratory combustion stability tests. The Design Unit took the design objectives from the Combustion Stability Committee and produced the hardware necessary for test and evaluation.

Rocketdyne initiated a program using a two-dimensional transparent combustion chamber running at low pressure in order to study droplet theories and performance, test baffle length and placement, and study angled baffles at various places on the injector's face.

After evaluating all the available data, the Committee specified six requirements for combustion stability in the F-1: (1) no leaks anywhere in the propellant system, (2) the feed system must be isolated from the combustion process, (3) combustion must be isolated radially, (4) control axial distribution, (5) eliminate random ignition locations and, (6) control chemistry.

The transparent combustion chamber idea had proven its use in the Atlas, Jupiter and Thor programs, as well as the ongoing H-1 program. The test program for the H-1 was also used in support of the F-1 program. These efforts helped to determine that large orifices for the RP-1 fuel coupled with relatively high combined propellant velocity contributed to dynamic stability; that the outer zone of the injector, near the walls of the thrust chamber, were sensitive to wave amplification; and that it would be necessary to install baffles to achieve dynamic stability.

Although all systems of the F-1 were studied as potential sources of combustion instability, the primary effort focused on the injector as the key element to achieve dynamic stability. Three injector designs were determined to have the most promise for further development of this engine: these being known as the 5U Unbaffled, 5U Baffled and double-row baffled units. Even so, in tests all three showed spontaneous instabilities. The 5U had the best performance characteristics. Its performance would be diminished slightly by the addition of baffles, but these proved to be essential for combustion stability and a dozen injector baffle designs were analyzed. Rocketdyne also performed hydraulic propellant studies from the pumps to the injector, leading to modifications which included injector dome torus dam inserts, conical plates, fuel port inserts and isolation tabs, flow dividers, inner and outer oxidizer baffles and oxidizer feed systems dams. These revisions had a significant and favorable effect in suppressing the occurrences of self-triggered instability. It was also determined that relocating the combustion zone of the propellants further away from the injector face contributed to combustion stability. All these efforts were focused on determining the Preliminary Flight Rating Test (PFRT) injector configuration, the final selection of which, based on the 5U design, was made in June 1963. Rocketdyne then set out to develop the Flight Rating Test (FRT) injector. This focused on further improving dynamic stability characteristics, efficiency and engine performance. It involved refining the propellant orifice design, including orifice angle, establishing optimum propellant velocity, injector near-wall modifications and other development work. By June 1964, based on successful development testing, the basic design features of the FRT injector had been established. By the end of the year, the FRT had achieved the required performance, dynamic stability, combustion instability damping and the other specified engine parameters. The next major goal was the Flight Qualification Injector. This evolved from component testing conducted between January 1965 and September 1966. These studies produced component refinements which resulted in the completion of injector stability demonstrations qualification by November 1965. One of the stipulations of this milestone was the damping out within 45 milliseconds

2-15 A mockup of the F-1 engine on the assembly floor in 1967. (Rocketdyne, Harold C. Hall Collection)

of deliberately induced instability. All these injector refinements were validated in mainstage testing at the Edwards Field Laboratory. The successful completion of Project First to resolve F-1 combustion instability was concurrent with other areas of engine development; particularly that of the turbopump.

## TURBOPUMP DEVELOPMENT

There were a number of LOX pump failures during F-1 component and engine tests, most of which resulted in explosions. During a test at MSFC on November 18, 1963 to determine the outcome of a premature shutdown by deliberately closing the LOX prevalve, there was a turbopump failure. As Bob Biggs recalled, this test proved to be fortuitous.

> The test in November 1963 was an extremely lucky one. One of six LOX pump impeller vanes broke completely loose, turned in the flow field and wedged between the two adjacent vanes. The broken vane cause a flow blockage which resulted in a loss of performance, but it continued to run without causing a fire or explosion. The failed hardware provided good information on the potential for vane fatigue failures. Design changes were made to strengthen the vane, but the program suffered three more impeller vane fatigue failures that resulted in explosions. The last one was in December 1965. The pump had been tested for a total of 5,070 seconds with additional vane improvements. No flight pump ever exceeded 800 seconds, so a time limit was established at 3,500 seconds which provided adequate margin against a fatigue failure.
>
> The four vane failures occurred in a run time clustered around 110 seconds. There was great concern that there was some significance to that time. It was called the 110 second phenomenon, and exhaustive investigation concluded that the failure times were a freak coincidence. In November 1967, I was at KSC in Firing Room Two to watch the first flight of the Saturn V. I and others familiar with the F-1 program were apprehensive until after 110 seconds had gone by, before we could really enjoy the spectacle. It remains the 110 second mystery.

These turbopump problems were resolved by refining component clearances and tolerances, shot-peening the impeller blade edges, coating metal mating surfaces with molybdenum, and realistically limiting the test life of the LOX impellers.

## OTHER COMPONENT FAILURES AND SOLUTIONS

To gain the performance and reliability required of the F-1 engine, other component failures of lesser magnitude had also to be addressed. Among the exotic materials used was Rene 41 for the turbine manifold. At that time, its use in rocket engines was new. In manufacturing the turbine manifold, cracks were found near sites where welding was required. Resolving this problem was achieved by proper component

2-16 Blankets protected the F-1 engine both at ignition and against base heating in the upper atmospheric phase of the S-IC's powered flight. (Rocketdyne, Harold C. Hall Collection)

preparation and specific training for welding the turbine manifold. The introduction of automatic welding later eliminated this variable from component to component. Another problem was injector ring–land separation resulting from increased testing time on research and development engines. This was discovered after a number of production engines were delivered to NASA. It was the result of a less than perfect bond of the braze material between the injector ring and the land. Subsequent lands were gold-plated to provide a better bonding surface for the braze material, and tests confirmed that this was sufficient.

There were F-1 engine flight pressure and temperature transducer failures during development. These transducers were essential to provide readings on the engine's performance and condition in flight. The highest failure rates occurred in the most severe operating conditions in the engine, and the lost transducers included those for thrust chamber pressure and turbine outlet pressure. These failure were usually due to contamination within the transducer itself or to the vibration damaging a capacitor or resistor. Rocketdyne initiated a development program to improve the durability and reliability of both types of transducers. They were redesigned and the electronic components were selected for improved endurance. High vibration loads in testing were also found to induce cracks around welded joints of the hydraulic control lines carrying RP-1 propellant. Although no outright failures occurred, this issue had to be resolved to rule out a catastrophic failure and resulting fire. The solution required only additional and better supporting brackets and paying more attention to control line alignment.

## ENGINE THERMAL INSULATION

A critical aspect of the F-1 engine was its thermal insulation. Most static tests were conducted without insulation, and there is a popular misperception that the Saturn V was launched with its F-1 engines similarly configured. However, close-up videos of the base of the vehicle at liftoff show that each F-1 engine was completely insulated. In fact, the problem started on the pad. The engines would be tremendously heated during the ignition sequence, as the base of the Saturn V was engulfed in flames and smoke prior to mainstage and liftoff. In flight, base heating was due to the curious effect of heat and flame propagation whereby, as the vehicle climbed through the ever-thinning atmosphere, the engine exhaust plumes expanded to occupy the lower pressure area directly at the base of the S-IC stage, causing hot exhaust actually to circulate around the engines themselves. To protect the F-1's components from this tremendous heat, exterior thermal insulation was required. Thermal insulation of the engines was initially Boeing's task, as part of the booster design, but owing to the surface complexity and interfaces of the F-1 engines with the S-IC, Boeing devolved the task to Rocketdyne.

A target weight of 150 pounds was allotted for insulating each engine, but this soon proved grossly optimistic. The weight of the thermal insulation system, which included all the support brackets and securing hardware, eventually reached almost

2-17 The liquid rocket engines used in the Saturn I, Saturn IB and Saturn V launch vehicles. (NASA/MSFC)

1,200 pounds. The thermal panels comprised Inconel foil of 0.152 mm thickness for the outer skin and 0.101 mm for the inner skin, with 1.0 cm of refrasil insulation batting in between. There were also laminated, aluminum-coated, asbestos blankets laced to attachment studs. The panels were form-fitting, and were capable of being removed and replaced at the Kennedy Space Center in order to permit servicing and maintenance of the engines. To simulate flight temperature conditions, the thermal insulation was tested using jet engine exhaust directed onto a thermal cocoon inside a controlled chamber at MSFC. These tests proved that the F-1 engine components would be adequately protected against damage by heat. However, the design of the thermal insulation system was not yet complete. While the 500-F test vehicle was at Launch Complex 39, a severe rainstorm soaked the thermal insulation of the F-1 engines and the refrasil insulation absorbed sufficient water to raise concerns that if a Saturn V were to be launched in this condition then superheated steam would be produced during engine operation and cause the thermal panels to rupture and rip off in flight. Tests conducted by Boeing with similarly soaked thermal insulation panels on engines at the contractor's plant showed the requirement for strategically located vents to relieve pressure and allow steam to safely escape in the event of the thermal panels absorbing moisture by any means.

## CONTINUING DEVELOPMENT AND IMPROVEMENT

Even after the F-1 engine was flight qualified, the development program continued in order to improve its components and demonstrate its reliability. The development of the uprated F-1A variant is covered in Chapter 9.

# 3

# F-1 engine project management by MSFC

The Rocketdyne F-1 engine project was transferred from the Air Force Rocket Propulsion Laboratory to NASA in 1958. On December 12, 1958 NASA announced the selection of Rocketdyne to develop a single-chamber rocket engine powered by RP-1 and LOX delivering 1 to 1.5 million pounds of thrust. The announcement was a formality, as the company had been at work on the F-1 engine for the Air Force for several years. NASA's initial management of the project was out of Headquarters under Adelbert O. Tischler and Oscar Bessio. Wernher von Braun's engineers were asked to provide technical support and attend meetings with the Air Force. Tischler and Bessio were involved in defining the engine's specifications to match NASA's perceived requirements, which were to power the Nova launch vehicle that was being considered for a mission to the Moon by a direct ascent trajectory. On January 19, 1959 NASA and Rocketdyne signed the formal development contract valued at $102 million. As soon as NASA acquired the Marshall Space Flight Center in July 1960, a process was begun to transfer the F-1 project management to MSFC, where the large Saturn and Nova boosters were going to be developed, and by November the process was complete.

Saverio "Sonny" Morea was assigned to manage the F-1 engine project at MSFC. After gaining a Bachelor's degree in Mechanical Engineering from City College of New York in 1954, Morea was recruited by North American Aviation to work in its Aerophysics Department on wind tunnel testing of the Navaho. The following year he was called to active duty and, as a result of his ROTC training while in college, reported to Aberdeen Proving Grounds in Maryland as a Second Lieutenant. Four months later, he was sent to the Redstone Arsenal, Huntsville, Alabama. On June 10, 1957, Morea, his wife and young child arrived in Huntsville. He reported to Gen. Holger Toftoy, who was in charge of the Army Ordnance Missile Command based there. Gen. Toftoy told him to report to Building No. 4888 to be interviewed by Dr. Wernher von Braun. As Morea explained to this author in 2007, Von Braun put him at ease:

3-1 Dr. Wernher von Braun, the first Director of the Marshall Space Flight Center, was admired as an excellent engineer and a superb manager. He promoted the concept of individual automatic responsibility of his managers and engineers. (NASA/MSFC, Frank Stewart Collection)

> His words to me were, 'I see that you are a mechanical engineer. We do a lot of work here requiring your background. What I would like you to do is spend the next six to seven weeks touring the different laboratories, and talk to the various heads of the laboratories and engineering departments as to where you might feel comfortable working. Come back to me and I'll assign you there.'

Morea joined the Structures and Propulsion Laboratory, working in the area of thermodynamics for Hans Paul, who was one of the original German scientists who came to the U.S. with von Braun in 1945. However, shortly thereafter Morea was reassigned to the Technical Liaison Office, where he served as a liaison engineer on the Redstone's thrust vector control system; many of the components for which were being developed by the Ford Instrument Company in New York. As Morea recalled:

> At the end of my two year military commitment, Dr. von Braun offered me a job to stay in the technical liaison office in Huntsville. I told him I would like to get back into propulsion. That was in Konrad Dannenberg's office. The first job Konrad gave me was the Jupiter S-3D engine.

The Saturn I project followed, and Morea's duties included propulsion studies to determine the number of engines and the best engine configuration for this powerful booster. When the F-1 engine was given to MSFC, the Technical Liaison Office was renamed the Propulsion Program Office with Leland Belew in charge. Von Braun asked Belew to recommend individuals within his office to manage the F-1 engine project, the H-1 engine project and the J-2 engine project. Belew evaluated his staff by their performance, and recommended Morea as the F-1 engine project manager,

Jack Seemore as the H-1 engine project manager and Bud Drummond as the J-2 engine project manager. Morea recalled his first actions:

> What I asked for immediately, was to have a representative from my office on site at Rocketdyne at all times during the development of the F-1. That gentleman was Frank Boffola. Rocketdyne had an office at MSFC and they had Vince Wheelock as the resident field manager. Rocketdyne did all the design on the F-1 but they were strongly encouraged by the involvement of knowledgeable 'hands on' engineers at Marshall as to approaches they ought to consider in design changes. We were on the critical path for the whole Apollo program. A day's slippage in the F-1 engine program would have resulted in a day's slippage in the Apollo program.

## MANAGING COMBUSTION INSTABILITY

Almost every liquid rocket engine that Rocketdyne developed displayed combustion instability to some degree, each uniquely, and eliminating it had required devising a variety of remedies. Morea was familiar with the combustion instability suffered by the E-1 engine, which took nearly two years to resolve. Unfortunately, steps taken to resolve combustion instability in one engine did not necessarily directly transfer to another engine. Brief tests of the F-1 in 1961 and early 1962 displayed combustion instability so severe that the Rough Combustion Cutoff sensors routinely triggered within seconds of engine start up, to prevent its destruction. When rough combustion persisted well into 1962, MSFC joined forces with Rocketdyne to resolve the issue. In the fall of 1962 Morea formed an ad hoc committee at MSFC that included some of the nation's best minds as consultants: Dr. Richard Priem of the Lewis Research Center, Robert Richmond of MSFC, Professor Luigi Crocco and David T. Harrje of Princeton University, Donald Bartz, Max Clayton and Jack Rupe of JPL, Prof. R.J. Osborn of Purdue University, Prof. E.S. Starkman of the University of California at Berkeley, and even Fred Reardon and Herbert B. Ellis from rival Aerojet General. As Morea recalls:

> We asked Jerry Thompson at MSFC to head up this team and work with a Rocketdyne team to put together a program to understand this phenomenon and find a solution to the problem. I gave them a budget, with few constraints, to use as they saw fit to solve the problem. I gave them a blank check, is what it amounted to. And I gave them all the authority that went with my own position as project manager to get something done and do it quickly. They took a 'shotgun' approach [exploring several approaches in parallel] because at that time we didn't have the software programs we have today to simulate the combustion process. We put this team together to work with the Rocketdyne team headed up by Paul Castenholz, who was a brilliant engineer. There was a strong feeling early on, based on our experience with the H-1 engine, that [the injector face] would require baffles in order to attenuate this high-frequency oscillation that was going across

3-2 Leland F. Belew was the first Manager of the Engine Program Office at the Marshall Space Flight Center. William D. Brown became Manager of this Office in the mid-1960s. To Belew's left is Dr. George Mueller. (University of Alabama in Huntsville Collection)

> the injector's face. Basically, it was an empirical solution that we found, based on our knowledge of what we were able to do with the H-1 engine.

In terms of F-1 engine project management, resolving the issues of combustion instability were of the highest priority for Leland Belew. In an interview with Roger E. Bilstein years later, Belew reflected upon the scale of the task:

> The actual development of an engine has to consider a lot of things like the flow [of the fuel and oxidizer], cooling characteristics, combustibility, and susceptibility of the propellants to rough combustion. [...] The F-1, in the early years of its development, I certainly developed a healthy respect for its capability to act up in the area of rough combustion. We put a major effort in solving rough combustion problems and [...] developed techniques that I feel have contributed greatly to that technology. We took steps in that development cycle that hadn't been done before, and did it in such a way that we were able to develop models that have and will continue to contribute to any similar kind of engine development in the future. It is not by chance that you get that kind of system built. It is by applying a heck of a lot of know-how that's been accumulated through past experiences, and it is your ability to bring that to bear. [...] That really gets down to the confidence gained in that field that we continue to have such vehicles as [the Saturn V for] Apollo, which is very, very difficult to come by. One thing one has to always remember is you should not lose that confidence.

## CONTRACT NEGOTIATIONS AND RESPONSIBILITIES

A critical part of the management of many government projects is the procurement process; especially if the contract is for design and development, testing, production, acceptance and fielding of something that has not existed previously. This aspect of a project, and the dynamics which it creates, is seldom addressed. It is through the preparation of the Request for Proposal (RFP) by the government, the submission of a proposal by an industrial contractor and finally the negotiation of the contracts that the programmatic, technical, business and legal aspects of a program or project are drawn together to define the deliverables, the terms and conditions under which the work will be done, and the costs and fees that the government will pay. If the project is to succeed, this process places a premium not only on project management, but also on contract management. In the absence of a proper contract and an appropriate relationship between the government team and the contractor team, a program such as the F-1 engine development would be severely encumbered, and the government would probably not get what it had desired, or at least not in a timely manner. In this case, NASA needed an engine of unprecedented power, the development of which would be difficult but had to be accomplished to meet a deadline. Rocketdyne was the sole-source contractor for the F-1 engine. The contracted work was implemented by two basic contracts: one for design, development and testing of the engine, and the other for production, acceptance, fielding and flight support of the real engines. In an effort such as this, the unknowns, complexities and interdependencies involved impose a symbiotic relationship between the government and the contractor. On the other hand, the government is the customer, and must maintain proper oversight and penetration of the work to insure that the need will be met in an effective and timely manner. This working arrangement might be characterized as a kind of arm's length partnership in which the customer (the government) retains the ability to penetrate, question and critically evaluate the work being done by the contractor whilst the two worked together as a team.

Richard Brown, a veteran of the Engine Program Office, joined MSFC shortly after its creation. Prior to going to work for Sonny Morea and Frank Stewart in the F-1 Engine Project Office, he worked in the Liquid Power Plant Engineering and Propulsion Performance Evaluation Branches. He was assigned to the F-1 project in mid-1964, shortly after the majority of the combustion instability problems had been resolved. Brown became chief of development on the F-1 engine at MSFC, and also did significant work on the production contract. He recalled:

> This arm's length partnership arrangement produced a healthy dynamic that was critical to the success [of the project]. In the engine world, Leland Belew, Manager of the Engine Program Office, deserves much of the credit for evolving a wholesome arm's length partnership for the Saturn V engines. He was very insightful, and provided invaluable guidance and senior management leadership to the Saturn V engine project offices, of which the F-1 engine was a part. Within the F-1 Engine Project Office, Sonny Morea, the F-1 Project Manager, and Frank

3-3 Saverio "Sonny" Morea (left) was F-1 Engine Project Manager at the Marshall Space Flight Center from the start of the program until September 1966, at which time he was promoted to Deputy Manager of the Engine Program Office. In 1969 he received the NASA Exceptional Service medal from Administrator Dr. Thomas O. Paine. (NASA/MSFC, Sonny Morea Collection)

> Stewart, his deputy, provided excellent leadership that effectively implemented the arm's length partnership concept.

The F-1 engine project spanned more than a decade of management by MSFC. It had a broad scope that included design, development, production, fielding and flight support of the engine by Rocketdyne. Testing was done by both the government and the contractor. In the division of responsibilities, Rocketdyne did the development, qualification and acceptance testing and the government undertook exploratory and confirmation testing, and tested the engine in the S-IC stage.

As Brown recounts:

> Establishing good contracts for the kind of work needed to bring the F-1 into existence required a lot of hard work by a multidisciplinary team of project, technical, contracts, pricing; supported by the Defense Contract Audit Agency and legal people.

In Brown's opinion, the team he was privileged to work with in the development and negotiation of the various F-1 engine contracts was second to none. As he told this author:

> We had a very competent, professional team, and excellent expertise in all areas. Sonny Morea and Frank Stewart provided overall project leadership. Ron Bledsoe and Carlyle Smith, the F-1 leads in the Propulsion Division at MSFC, coordinated the technical effort. Ed Mintz was the F-1 quality engineer. E.B.

Craig was Manager of the Engine Contracts Branch in the Procurement Office; to whom Contracting Officer Tom Burton and contract administrators Al Jollif and Mike Burns reported—if I remember correctly. Herb Kithchens was the F-1 price analyst. Various procurement attorneys were involved, depending on the time and question involved. All were 'top drawer' folks who really knew their business and were dedicated to the mission. I was the lead for preparing the Statement of Work and supporting documentation, and also as the focal point for the project office for the overall effort. It was a great team.

Just after I got to the F-1 Project Office, we had a major effort to update the contract because we now knew better the extent of the work which would be required to make the needed design modifications to the engine (the Block II design), do the remaining development work and qualify the engine for flight. Subsequently, we had to make major revisions to the production contract as well. I found out in a hurry that the contracting part of the business was as complicated (maybe more so) than the engine itself. There is the Statement of Work prepared by the project office, supported by the technical people. Then there are the contract Terms and Conditions which are done by the contracts people with input and review by the legal people. There are so many things required by law and the procurement regulations [for the government] that it is almost mind boggling. The Terms and Conditions include government policy clauses—like the Buy America Act and the use of minority and disadvantaged businesses. Then there are all the [...] ground rules for reimbursement (these were cost reimbursement contracts), reporting requirements, fee arrangements and amounts, etc. On several occasions it was necessary to, shall we say, be 'innovative' to meet [...] all the constraints. Needless to say, these times were a real challenge to the contracts and legal people, and required some spirited discussion within the government team. The government team strove to make sure the deliverables were clearly defined in the contract: hardware, software, documentation, supporting analyses, progress reports, financial information, certifications, support efforts, etc.

In the F-1 Project Office we spent a lot of time on the Statement of Work for the Request for Proposal, and what the specifications would be that would accompany the Statement of Work. One key document that specified the end item we wanted (basically the performance, size and weight of the engine, and its operating parameters) was the Model Specification, or 'Mod Spec'. When I moved into the Project Office, it was apparent the Mod Spec was out of date, so Ron Bledsoe, Carlyle Smith and I updated it to what was needed at that point in time—1964. We presented it to Rocketdyne, and worked it with them, iterating and negotiating to get a spec that reflected what we thought was needed and what was possible. This is key to seeing that they do the right job. This was basically a technical negotiation of the type that often preceded the formal contract negotiations. [...] Rocketdyne and NASA were both skilled in such negotiations. On the F-1 project, the government was usually represented by the Deputy Project Manager, the F-1 engine Contracting Officer and I. The contractor was usually represented by their project manager or his deputy, the contracts manager, and

3-4 Francis "Frank" Stewart became F-1 Engine Project Manager in September 1966, following Morea's promotion. He is shown receiving the NASA Exceptional Service medal from NASA Administrator Dr. James C. Fletcher (center) and Dr. George M. Low, Deputy Administrator. (NASA/MSFC, Frank Stewart Collection)

others as needed to explain or defend their proposal. NASA had very specific requirements, and Rocketdyne wanted to insure they had a reasonable chance of meeting them. Negotiation involved arriving at an agreement on each specific point in the contract. A lot of money was involved in fulfilling these contracts, but contract negotiation was about far more than the dollars: it involved dates for tests, dates for delivering engines, how many engines, and many other issues. Negotiations were invariably contentious, but this was also part of the process. NASA expected Rocketdyne would disagree with certain areas of a contract, and even protest certain terms and conditions and requirements. Rocketdyne, for its part, did not want to appear too eager to agree to each and every requirement that NASA laid down. In many instances Rocketdyne had legitimate concerns about its ability to meet certain terms in a contract. These had to be negotiated to the satisfaction of both sides. Although sometimes grueling, the negotiations with Rocketdyne were always interesting and on occasion even amusing.

Rocketdyne had a real sharp and competent contracts manager named Jim Owlsley. He could be a tough negotiator. One of his negotiation strategies was to explode his temper at judicious points of the contract negotiations. We used to refer to this his rain dance. We were getting ready to negotiate the contract after evaluating the proposal for the last phase of the F-1 engine development. We were up on the tenth floor of the 4200 Building at Marshall. During pre-negotiation discussion before Rocketyne came in, we developed our offer. The contracting officer remarked, 'We have the offer we'll put on the table, but of course we'll have to go through Jim Owlsley's rain dance before we can get around to discussion and negotiation.' So during the meeting with Rocketdyne and after [...] the government had put its offer on the table, Owlsley started his rain dance. 'That's one of the most ridiculous offers I've ever seen!' he cried, while pounding the table. At that moment, there was a big flash of lightning and a clap of thunder [...] and all of us on the government side died laughing. He couldn't figure out what was going on, and looked very perplexed. Then we let him in on our little joke. I have to say that Jim looked pretty sheepish at that point. From that point on, we had a very productive negotiation. Through the negotiation process, a mutual understanding was established between the two organizations as to what it was going to take to do the job. This is a very important part. It was where we negotiated the number of engines, the number of engine tests, the adoption of specifications negotiated between the technical people at Marshall and Rocketdyne—in short, every aspect of what was to be done and the terms and conditions under which it was to be done. When you evaluate a proposal for the work they are going to do, you get into how many hours they are going to take to build an engine, how many times and seconds [of duration] it is going to be tested, the schedule, the facility needs, the spares and scrap rate and so on. It is very detailed.

One of the major challenges [of the F-1 engine] was that the development work and the production work were being carried on in parallel owing to the tight schedule. The development contract was to do the development work and pass qualification—what today we would say was certifying it for flight. Once you solved a problem in development, you had to get that change incorporated into the production engines. [And, of course, all] of these changes had to be approved by the government before incorporation into flight hardware. That is where configuration management and change control was so important. You had to keep track of which changes were applicable to which flight engines, and which changes had been incorporated into which engines—remember we were doing all this without computers. An orderly way of evaluating changes and keeping track of them was a must.

Once the contract was negotiated, the contractor performed the work specified in the contract, with the government's team participating to minimize errors, undertake engineering analyses and do tests to verify the design and development work by the contractor and insure that it complied with the contract. The contractor had to draw up a great deal of documentation, which the government reviewed to confirm that it met what was required to fulfill the mission. The contract specified those things that required formal government approval. In this regard, Brown noted:

Dr. von Braun fostered the idea that in order to procure something effectively of a high technology nature, you needed to be an informed, smart buyer. So, the

3-5 Gen. Samuel Phillips visited the Rocketydne facilities in January 1964 to see F-1 and J-2 engine manufacturing firsthand. His Air Force program management skills were instrumental in the achievement of Apollo 11's lunar landing in July 1969, after which the Air Force made him Commander of the Space and Missiles Systems Organization. (NASA/KSC)

> engine group at Marshall [...] were responsible to oversee technically what Rocketdyne did. They had what von Braun called Automatic Responsibility to bring to light things they were concerned about from a technical standpoint. It was more than oversight, it was *insight*. We had to be capable of doing the engineering calculations, analyses, etc., to check on the work at Rocketdyne. The philosophy was much like a peer review process. I think this approach has a lot to do with our success under great schedule pressure.

Brown concluded by saying, "Well, what do you take away from all of this? It's that getting to the moon was ninety percent grunt work and ten percent glamour."

## BROAD PROJECT MANAGEMENT ASPECTS

The F-1 Engine Project Office at MSFC remained busy throughout the 1960s and early 1970s, until the launch of the Skylab Workshop. It continued to monitor F-1 engine flow from Rocketdyne to the Michoud Assembly Facility, and the S-IC stage test results at the Mississippi Test Facility. It is significant that after the combustion instability and turbopump problems were resolved, the F-1 engine did not encounter any other significant issues.

In September 1966 Severio Morea was promoted from being F-1 Engine Project Manager to the Deputy Manager of the Engine Program Office. Deputy F-1 Engine Project Manager Frank Stewart was in turn promoted to Project Manager, a position he held until the end of the F-1 engine project at MSFC. Morea was later selected to manage the Lunar Roving Vehicle program at MSFC. Leland Belew was promoted from the Engine Program Office to manage the Skylab program.

On the completion of the Apollo and Skylab programs, MSFC would find new challenges in the management of Space Shuttle Main Engines (SSME) and Solid Rocket Motors (SRM) as part of the Space Transportation System. Frank Stewart was selected to head the definition phase of the SSME, with Richard Brown as his deputy. Mr. Brown was subsequently selected to lead the preparation of the RFP for both the SSME and the SRM, and was heavily involved in their Evaluation Boards.

# 4

# F-1 engine description and operation

The F-1 rocket engine is a single-start, 1.5 million pound fixed-thrust, bi-propellant launch vehicle propulsion system. It uses liquid oxygen (LOX) as the oxidizer and RP-1 (refined kerosene) as the fuel, with a 2.27:1 mixture ratio of LOX to RP-1. The flow rates are 24,811 gallons per minute for LOX and 15,471 gallons per minute for RP-1. The propellants are combined and burned in the thrust chamber assembly. The burning gases at 5,970 degrees F are expelled through an expansion nozzle to create thrust, with a thrust chamber pressure of 965 psi. The engine is bell-shaped with an area expansion ratio of 16:1; this being the ratio of the area of the throat, just beyond the injector plate, to the aperture of the thrust chamber extension. All five of the F-1 engines in the S-IC stage of the Saturn V launch vehicle are identical, apart from the fact that the center engine is fixed whilst the four outboard engines can be gimbaled. The flight configuration weight of the F-1 is 18,500 pounds.

The major engine systems are the Propellant Tank Pressurization System, the Propellant Feed System, the Thrust Chamber Assembly, the Turbopump, the Gas Generator System, the Engine Interface Panel, the Electrical System, the Hydraulic System and the Flight Instrumentation System.

## PROPELLANT TANK PRESSURIZATION SYSTEM

The pressurization system heats gaseous oxygen (GOX) and helium to pressurize the propellant tanks. It includes a heat exchanger, a heat exchanger check valve, a LOX flowmeter, and various heat exchanger lines. The LOX source for the heat exchanger is tapped from the thrust chamber oxidizer dome, and the helium is drawn from cold storage helium tanks housed in the main LOX tank. LOX flows from the thrust chamber oxidizer dome through the heat exchanger check valve, LOX flowmeter, and the LOX line to the heat exchanger.

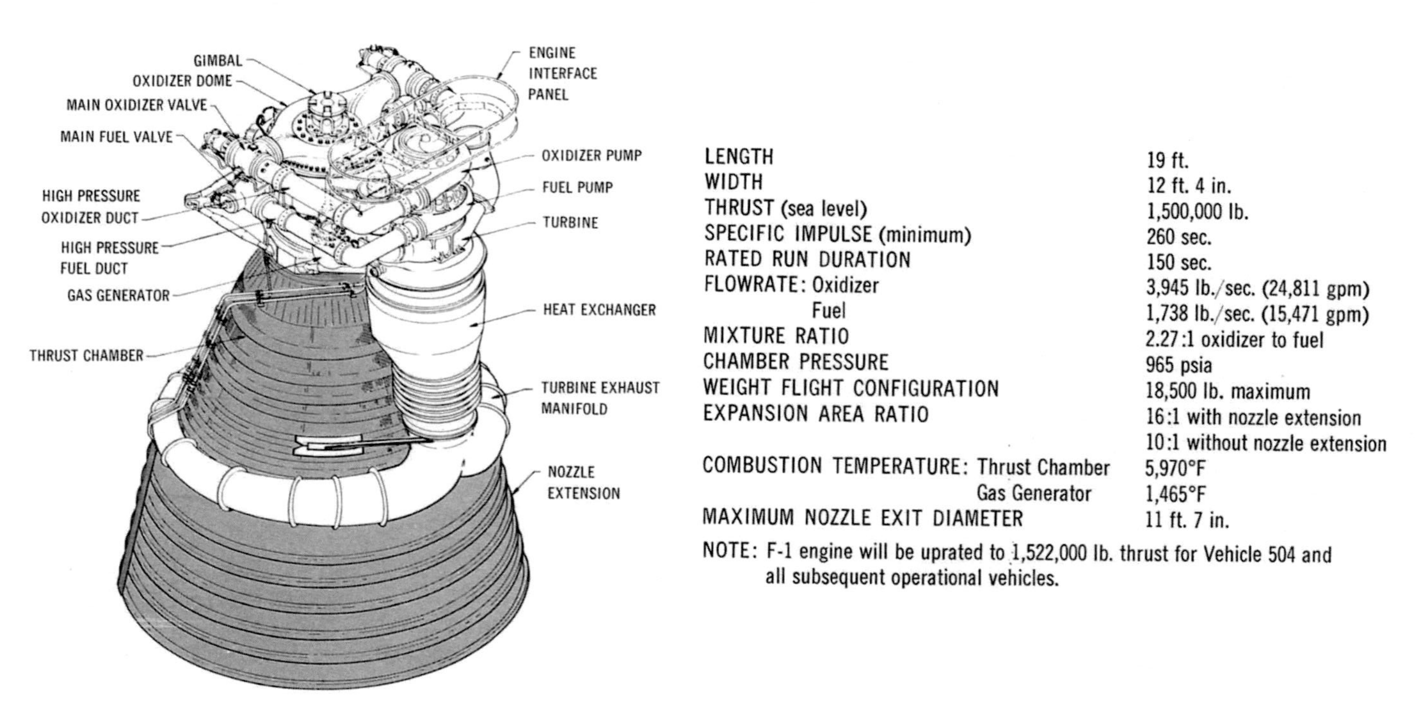


4-1 An illustration of the early production F-1 engine. Design of the propellant lines and nozzle extension would be the most notable changes in later engines. (NASA)

### Heat exchanger

The heat exchanger heats GOX and helium with hot turbine exhaust gases, which pass through the heat exchanger, whose duct has four oxidizer coils and two helium coils. It sits between the turbopump manifold outlet and the thrust chamber exhaust manifold outlet, and its shell contains a bellows assembly to compensate for thermal expansion during engine operation.

### Heat exchanger check valve

The heat exchanger check valve prevents GOX or vehicle prepressurizing gases from flowing into the oxidizer dome. It consists of a line assembly and a swing check valve assembly. It is installed between the thrust chamber oxidizer dome and the heat exchanger LOX inlet line.

### LOX flowmeter

The LOX flowmeter is a turbine-type volumetric liquid-flow transducer employing two pickup coils. Rotation of the LOX flowmeter turbine generates an alternating voltage at the output terminals of the pickup coils.

### Heat exchanger lines

LOX and helium are routed to and from the heat exchanger through flexible lines. The GOX and helium lines terminate at the vehicle connect interface. The LOX line connects the heat exchanger to the heat exchanger check valve.

## PROPELLANT FEED CONTROL SYSTEM

The propellant feed control system transfers LOX and fuel from the propellant tanks into the pumps, which discharge into the high pressure ducts leading to the gas generator and the thrust chamber. The system consists of two oxidizer valves, two fuel valves, a bearing coolant control valve, two oxidizer dome purge check valves, a gas generator and pump seal purge check valve, turbopump outlet lines, orifices, and lines between the components. High pressure fuel is supplied from the propellant feed system of the engine to the thrust vector control system provided by the vehicle contractor.

### Oxidizer valves

Two identical oxidizer valves, designated No. 1 and No. 2, control LOX flow from the turbopump to the thrust chamber oxidizer dome, and sequence the hydraulic fuel to the opening port of the gas generator valve. Each of the oxidizer valves is a hydraulically actuated, pressure balanced, poppet type, and contains a mechanically

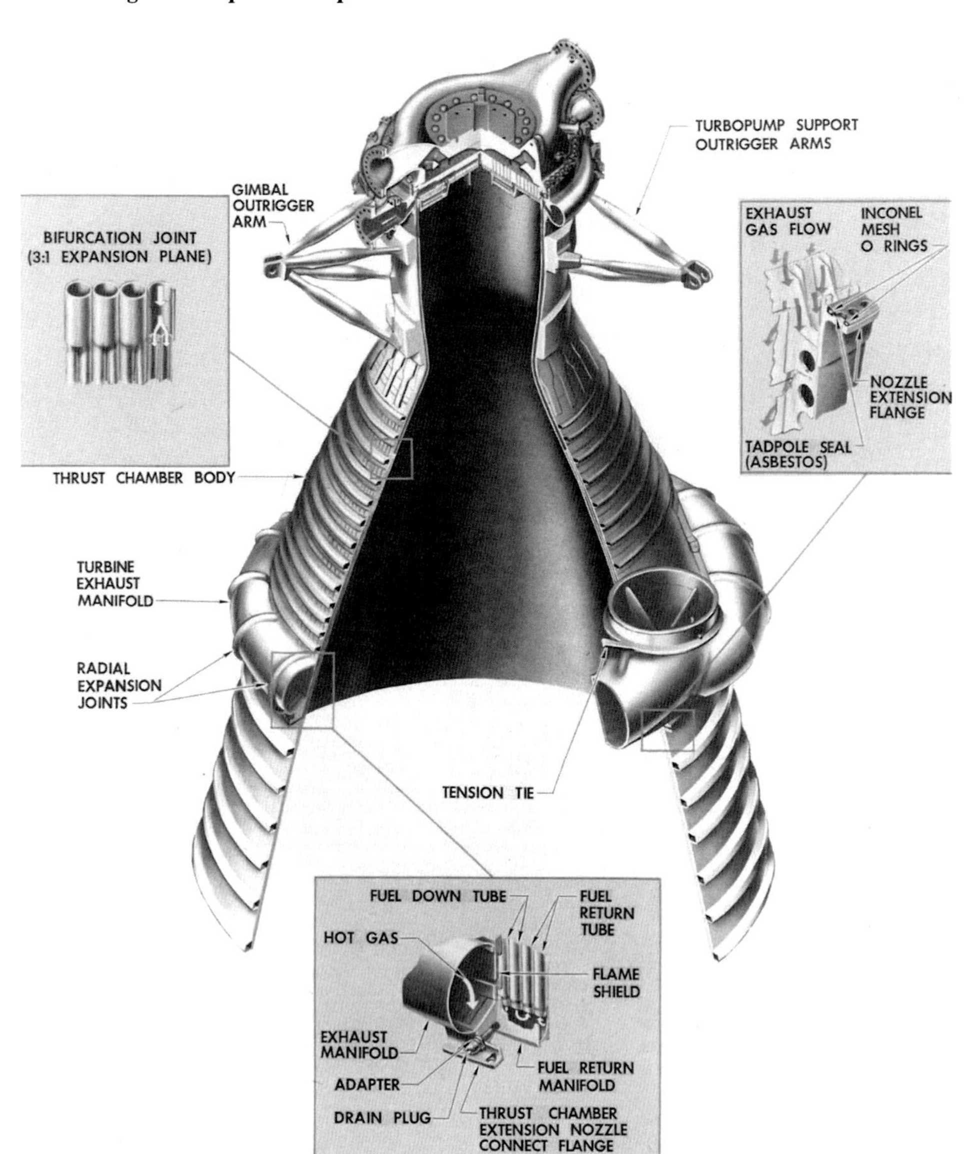


4-2 A cutaway of the F-1 showing the interface between the thrust chamber and the nozzle extension, and the bifurcation joints of the thrust chamber tubing. (Rocketdyne, Harold C. Hall collection)

actuated sequence valve. A spring loaded gate valve permits reverse flow for recirculation of the hydraulic fluid with the propellant valves in the closed position, but prevents fuel from passing through until the oxidizer valve is open 16.4 percent. As the oxidizer valve reaches this position, the piston shaft opens the gate, allowing fuel to flow through the sequence valve, which in turn opens the gas generator valve.

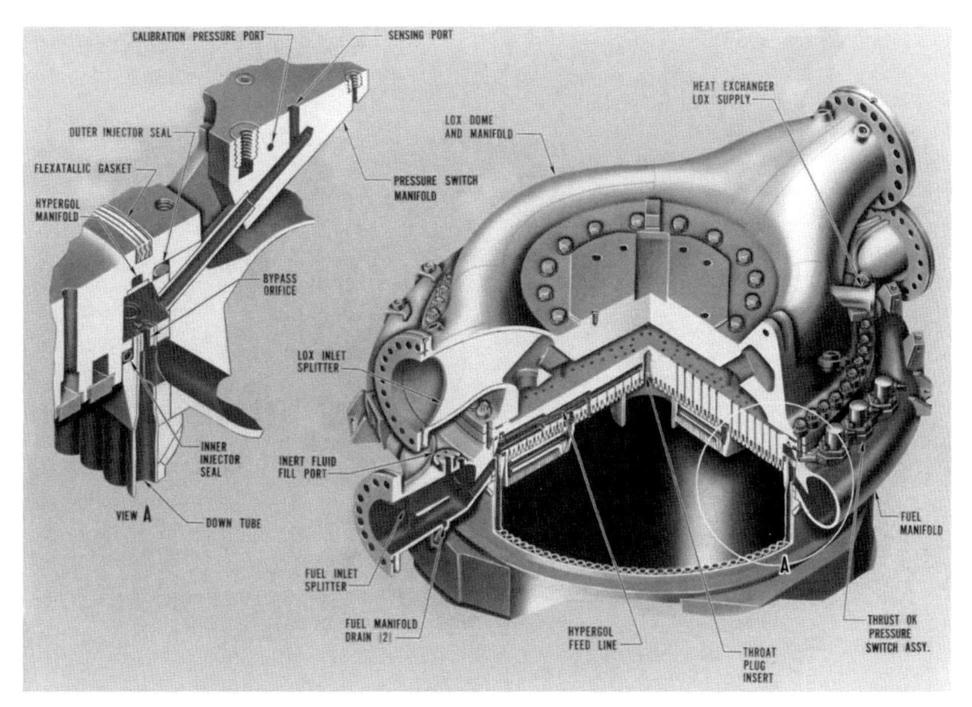


4-3 A cutaway of the F-1 engine's LOX dome assembly, propellant inlets and injector. (Rocketdyne, Harold C. Hall collection)

A position indicator provides relay logic in the engine electrical control circuit and instrumentation for recording movement of the valve poppet. The two oxidizer dome purge check valves, mounted on each of the oxidizer valves, allow purge gas to enter the oxidizer valves, but prevent oxidizer from entering the purge system.

## Fuel valves

Two identical fuel valves, designated No. 1 and No. 2, are mounted 180 degrees apart on the thrust chamber fuel inlet manifold, and control the flow of fuel from the turbopump to the thrust chamber. When the valves are in the open position at rated engine pressures and flow rates, they will not close if hydraulic fuel pressure is lost. Position indicators in the fuel valves provide relay logic in the engine electrical control circuit and instrumentation for recording movement of the valve poppets.

## Thrust-OK pressure switches

Three pressure switches on a single manifold which is located on the thrust chamber fuel manifold, sense fuel injection pressure. These thrust-OK pressure switches are

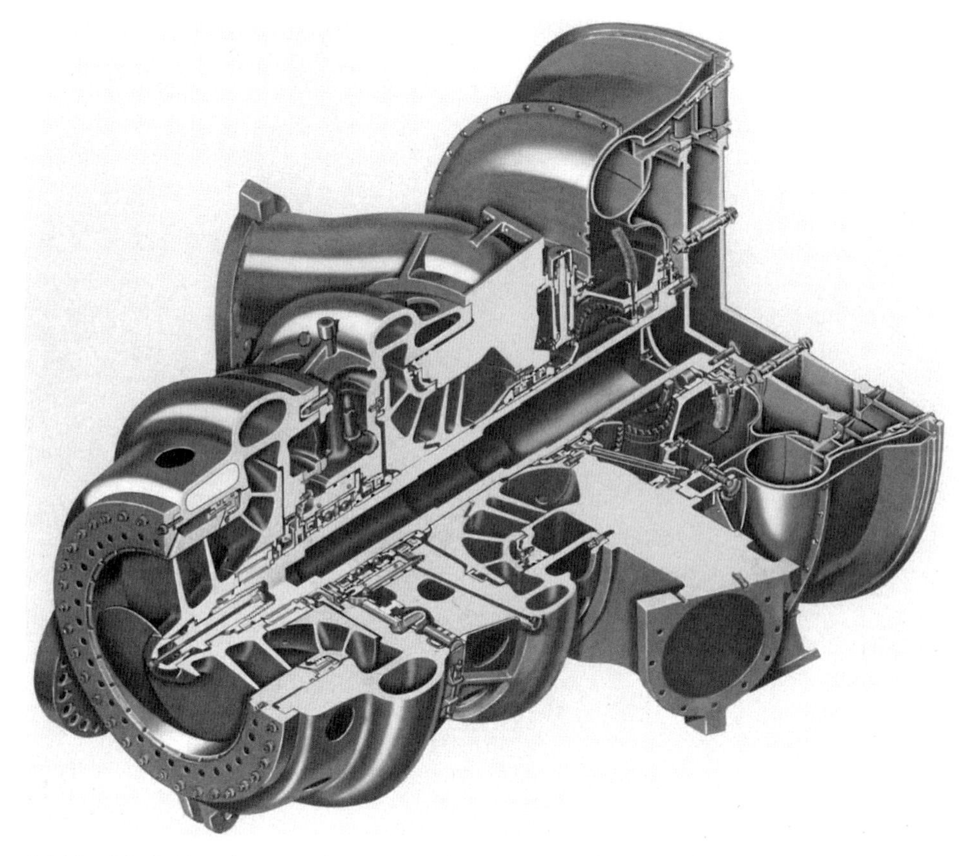

4-4 The Mark 10 turbopump incorporated a common drive shaft for (left to right) the LOX pump, fuel pump and turbine. (Rocketdyne, Harold C. Hall collection)

used redundantly in the vehicle to indicate that all five F-1 engines are operating satisfactorily. If the pressure in the fuel injection cavity falls, the switches deactuate, breaking the contact and interrupting the thrust-OK output signal.

## THRUST CHAMBER ASSEMBLY

The thrust chamber assembly includes the gimbal bearing, oxidizer dome, injector, thrust chamber body, thrust chamber nozzle extension and the thermal insulation. It receives propellants under pressure from the turbopump, mixes and burns them in a manner designed to expel the combustion gases at high velocity in order to produce thrust. The thrust chamber assembly also serves as the mounting point for all engine hardware.

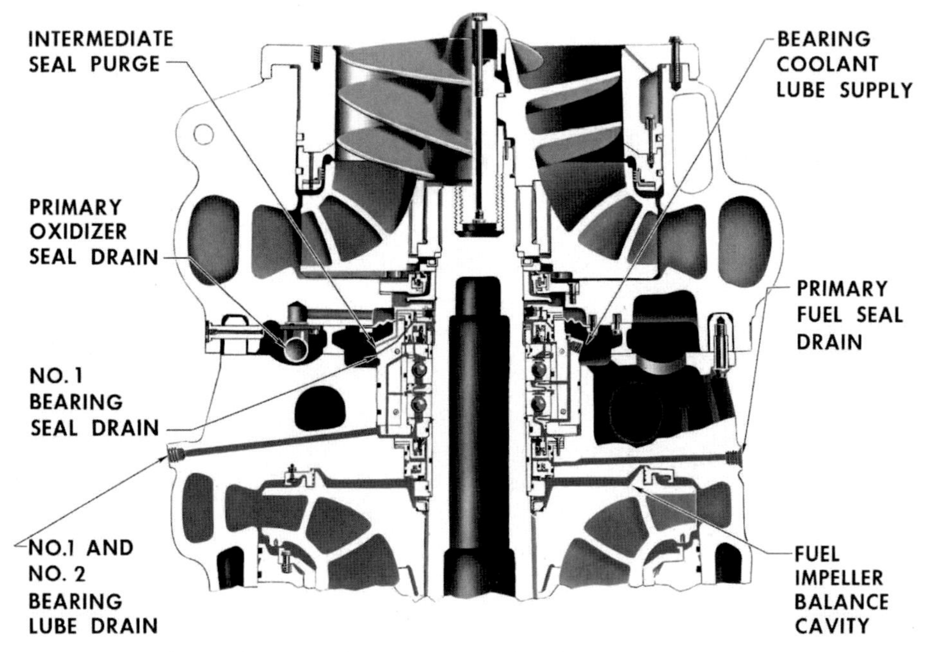


4-5 A cutaway of the turbopump showing the shaft bearings, LOX impeller and fuel passages. (Rocketdyne, Harold C. Hall collection)

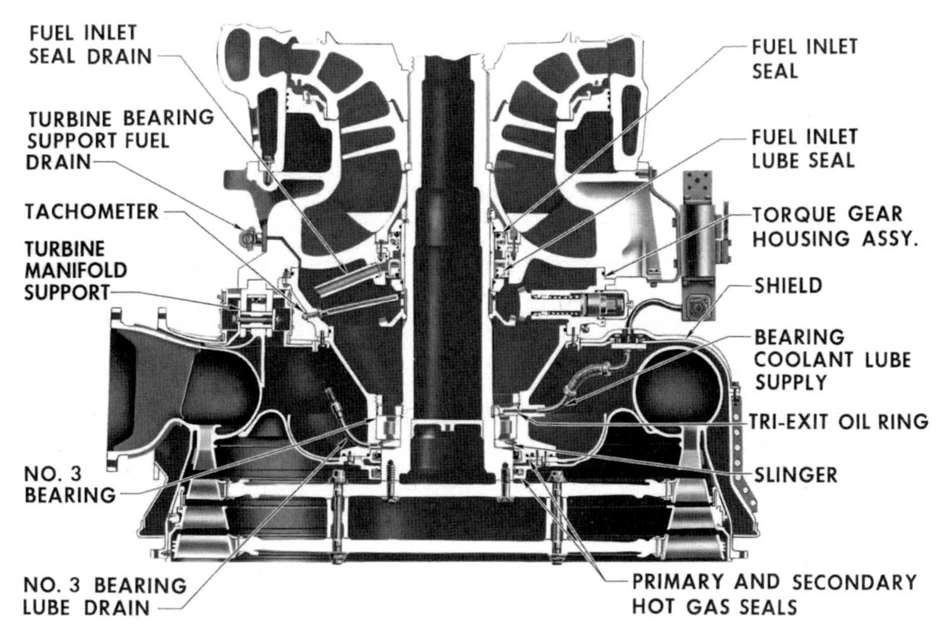


4-6 A cutaway of the lower turbopump showing the fuel passages and turbine manifold. (Rocketdyne, Harold C. Hall collection)

## Gimbal bearing

The gimbal bearing is attached to the oxidizer dome, and secures the thrust chamber assembly to the thrust structure of the S-IC stage of the Saturn V. The gimbal is a spherical universal joint that includes the socket-type bearing with a bonded Teflon-fiberglass insert that provides a low friction bearing surface. It permits a maximum pivotal movement of plus or minus six degrees in the [pitch and yaw] axes in order to facilitate thrust vector control. The gimbal transmits engine thrust to the S-IC and provides capability for positioning and thrust alignment.

## Oxidizer dome

The oxidizer dome functions as a manifold for distributing oxidizer to the thrust chamber injector, provides a mounting surface for the gimbal bearing, and transmits thrust forces to the S-IC thrust structure. Oxidizer enters the dome at a volume flow rate of 24,811 gallons per minute through two inlets set 180 degrees apart in order to maintain even distribution of the propellant.

## Thrust chamber injector

The thrust chamber injector directs propellants into the thrust chamber in a pattern that ensures satisfactory combustion. It is a multi-orificed structure with copper fuel rings and copper oxidizer rings forming its interior face, and contains the injector orifice pattern. Radial and circumferential copper baffles extend downward to divide the face into compartments. The baffles and rings, together with a segregated igniter fuel system, are installed in a stainless steel body. Fuel enters the injector from the thrust chamber fuel inlet manifold. In order to facilitate the engine start phase and reduce pressure losses, part of the flow of fuel is introduced directly into the thrust chamber, with the remainder (controlled by orifices) flowing through alternate tubes which run the length of the thrust chamber body to the nozzle exit, where it enters a return manifold and flows back to the injector through another set of tubes, thereby regeneratively cooling to the thrust chamber. Oxidizer enters the injector from the oxidizer dome.

## Thrust chamber body

The thrust chamber body provides a combustion chamber for burning propellants to generate pressure, and an expansion nozzle to expel the combustion gases at high velocity to produce thrust. The thrust chamber is tubular-walled and regeneratively cooled by fuel making its way to the injector, and the nozzle is bell-shaped. There are four sets of outrigger struts on the exterior of the chamber: two sets being for the turbopump, and the others being for the gimbal actuators provided by the vehicle contractor. The thrust chamber incorporates a turbine exhaust manifold at the nozzle

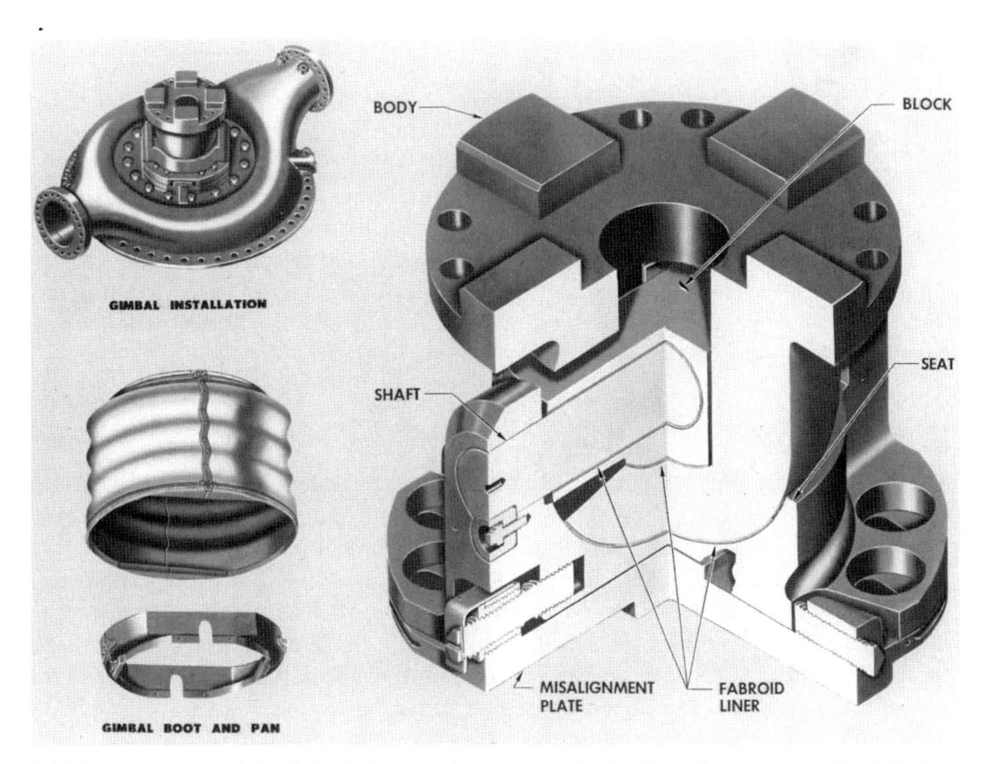


4-7 The gimbal assembly of the F-1 engine featured a spherical interface that permitted the four outboard engines to continue to transmit the entire thrust to the S-IC thrust structure even when canted. (Rocketdyne, Harold C. Hall collection)

exit, and a fuel inlet manifold at the injector end which directs fuel to the 'down' tubes. Brackets and studs welded to the reinforcing 'hatbands' surrounding the thrust chamber provide attachment points for thermal insulation blankets. Fuel enters the fuel inlet manifold through two inlets set 180 degrees apart. From the manifold, 70 percent of the fuel is diverted through 89 alternating corrosion-resistant steel 'down' tubes the length of the chamber. A manifold at the nozzle exit returns the fuel through the remaining 89 'up' tubes to supply the injector. The fuel flowing through the tubes provides regenerative cooling of the chamber walls during engine operation. The thrust chamber tubes are bifurcated. That is, they are comprised of a primary tube from the fuel manifold to the 3:1 expansion ratio area. At that point, two secondary tubes are spliced into each primary tube. This maintains the desired cross-sectional area in each tube through the large-diameter belled nozzle section. The turbine exhaust manifold, which is fabricated from preformed sheet metal shells, and forms a torus around the aft end of the thrust chamber body, receives turbine exhaust gases from the heat exchanger. Upon entering the manifold, gases are distributed uniformly. As the gases are expelled from the manifold, flow vanes in the exit slots provide uniform static pressure distribution in the nozzle extension. Radial expansion joints compensate for thermal growth of the manifold.

### Thrust chamber nozzle extension

The thrust chamber nozzle extension increases the expansion ratio of the thrust chamber from 10:1 to 16:1. It is a detachable unit made of high strength stainless steel, and is bolted to the exit end ring of the thrust chamber. Its interior is protected from the 5,800 degrees F engine exhaust gas environment by film cooling using the 1,200 degrees F turbine exhaust gas. This enters the extension between a continuous outer wall and a shingled inner wall, passes out through injection slots between the shingles, and flows downward across the surfaces of the shingles to form a boundary layer between the inner wall of the nozzle extension and the hotter exhaust gases emerging from the main engine combustion chamber.

### Hypergol cartridge

The hypergol cartridge supplies the fluid to begin combustion in the thrust chamber. It is a cylinder with a burst diaphragm welded to each end, and contains a hypergolic fluid comprising 85 percent triethylborane and 15 percent triethylaluminum. As long as this fluid is in the hermetically sealed cartridge it is stable, but will spontaneously ignite upon coming into contact with oxygen in any form. In starting the engine, the increasing fuel pressure in the igniter fuel system ruptures the burst diaphragms, the fuel and the hypergolic fluid enter the chamber through a segregated system in the injector, and on coming into contact with the oxidizer already present the hypergolic fluid ignites, thereby initiating fuel combustion.

### Pyrotechnic igniter

Pyrotechnic igniters actuated by an electric spark provide the ignition source for the propellants in the gas generator, and re-ignite the fuel-rich turbine exhaust gases as they exit from the nozzle extension.

### Thermal insulation

The thermal insulation protects the engine from the extreme temperatures (2,550 degrees F, maximum) created by the radiant heat from the exhaust plume and from backflow of the exhaust into the engine compartment in flight. Two types of thermal insulators are used on the engine—foil-batting on complex surfaces, and asbestos blankets on large simple surfaces. They are made of lightweight materials, and are equipped with a variety of mounting provisions such as grommeted holes, clamps, threaded studs, and safety wire lacing studs.

## TURBOPUMP

The turbopump is a direct-drive unit comprising an oxidizer pump, a fuel pump and a turbine mounted on a common shaft. The turbopump delivers fuel and oxidizer to the

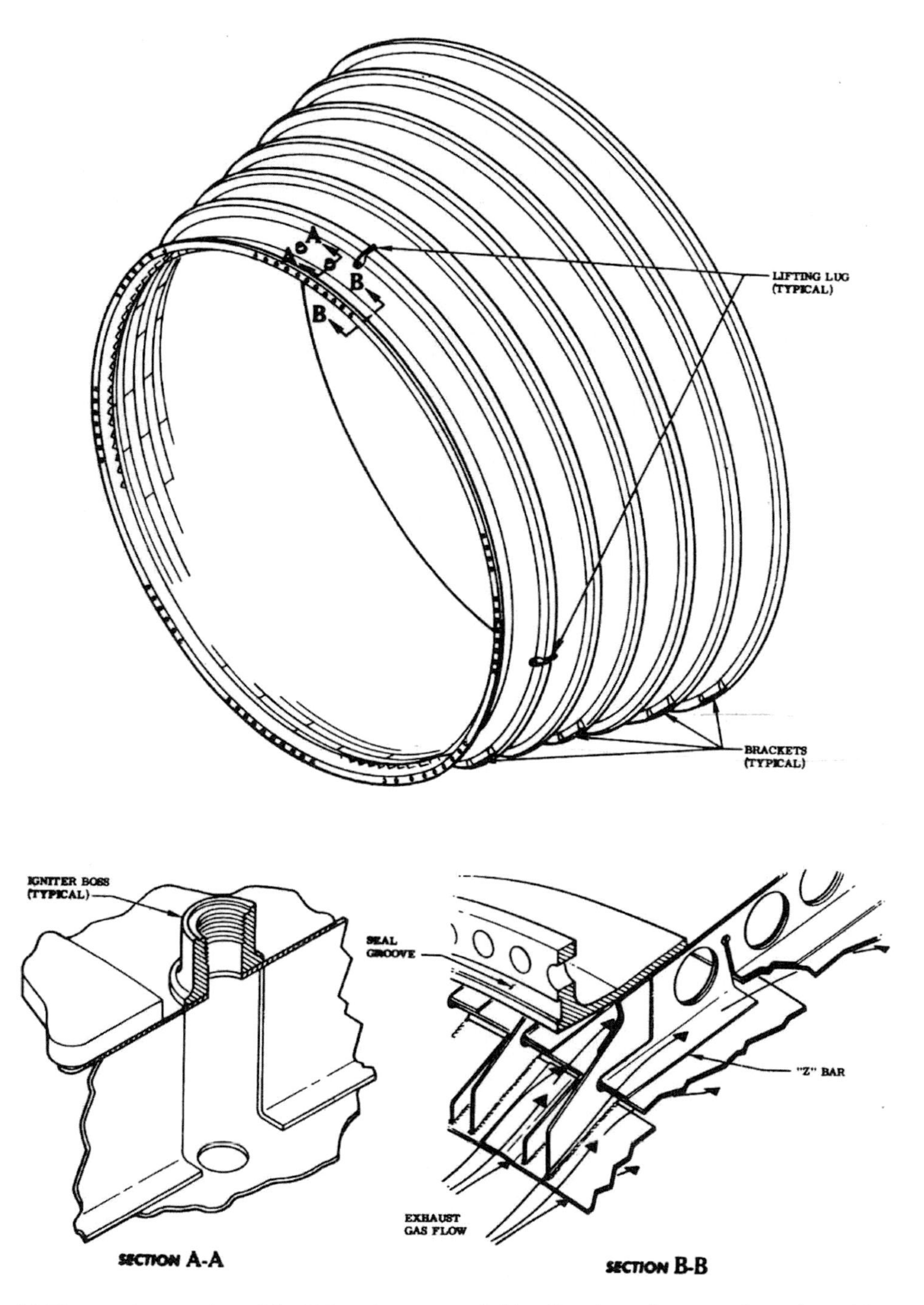


4-8 The nozzle extension of the F-1 engine was cooled by diverting exhaust gas from the turbine to passages on the inner walls in order to provide a boundary layer between the hot thrust chamber gases and the wall of the nozzle extension. (Rocketdyne, Vince Wheelock Collection)

gas generator and the thrust chamber. LOX enters through a single axial inlet in the shaft, and is discharged tangentially through dual outlets. Fuel enters through dual radial inlets, and is discharged tangentially through dual outlets. The dual inlet and outlet design provides a balance of radial loads in the pump. Three bearing sets support the shaft: No. 1 and No. 2 are matched tandem ball bearings in between the oxidizer and fuel pumps, and No.3 is a roller bearing between the fuel pump and the turbine wheel. A heater block provides the outer support for No. 1 and No. 2, and is used to prevent the bearings from freezing while chilling the oxidizer pump prior to engine start. During pump operation, the bearings are cooled with fuel. A gear ring installed on the shaft is used in conjunction with the torque gear housing for rotating the pump shaft by hand, and is also used in conjunction with a magnetic transducer for monitoring shaft speed. There are nine carbon seals in the turbopump: primary oxidizer seal; oxidizer intermediate seal; lube seal for No. 1 bearing; lube seal for No. 2 bearing; primary fuel seal; fuel inlet seal; fuel inlet oil seal; hot-gas primary; and hot-gas secondary seal. The main shaft and the parts attaching directly to it are dynamically balanced prior to its final installation into the turbopump.

## Oxidizer pump

The oxidizer pump supplies oxidizer to the gas generator and the thrust chamber at a fixed flow rate of 24,811 gallons per minute. It incorporates an inlet, an inducer, an impeller, a volute, bearings seals and spacers. Oxidizer is introduced into the pump through the inlet. The inducer in the inlet increases the pressure of the oxidizer as it enters the impeller, in order to prevent cavitation. The impeller accelerates the fluid to the desired pressure and discharges it through diametrically opposed outlets to the high pressure oxidizer lines to the gas generator and the thrust chamber. The inlet, which attaches to a duct from the main oxidizer tank, is bolted to the oxidizer volute. Two piston rings between the inlet and the volute expand and contract in response to temperature changes in order to maintain an effective seal between the high and low pressure sides of the inlet. Holes in the low pressure side of the inlet allow leakage past the ring seals to flow into the suction side of the inducer, thus maintaining a low pressure. The oxidizer volute is secured to the fuel volute with pins and bolts which prevent rotational and axial movement. The primary oxidizer seal and spacer located in the oxidizer volute prevent fuel from penetrating the primary oxidizer seal drain cavity. The oxidizer intermediate seal directs a purge flow into the primary seal and No. 3 drain cavities where the purge acts as a barrier to permit positive separation of the oxidizer and bearing lubricants.

## Fuel pump

The fuel pump supplies fuel to the gas generator and the thrust chamber at a flow rate of 15,471 gallons per minute. It has an inlet, an inducer, an impeller, a volute, bearings, seals and spacers. Fuel is introduced through the inlet. The inducer in the inlet increases the pressure of the fuel as it enters the impeller, in order to prevent cavitation. The impeller accelerates the fuel to the desired pressure and discharges it

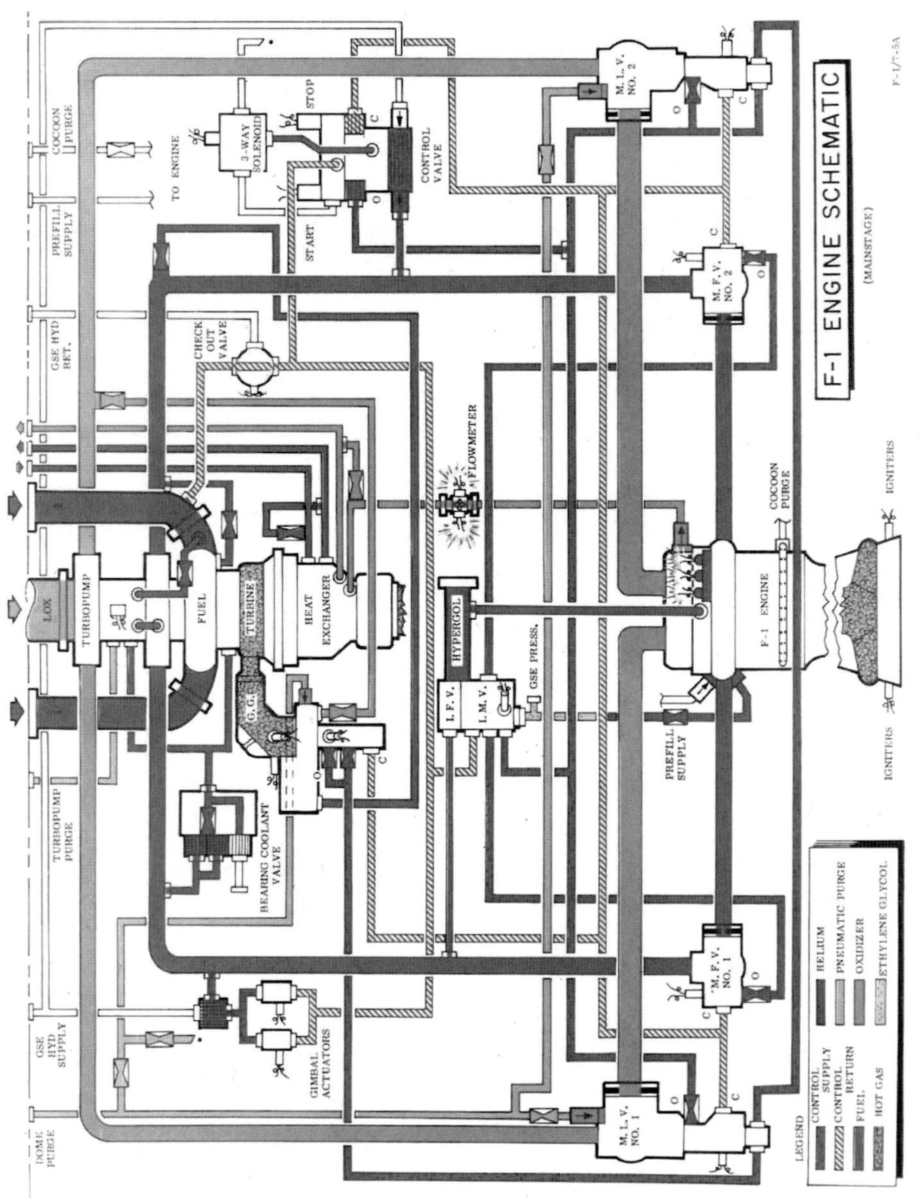


4-9 A schematic showing F-1 mainstage operation. (Rocketdyne, Harold C. Hall Collection)

through two diametrically opposed outlets into the high pressure fuel lines leading to the gas generator and thrust chamber. The fuel volute is bolted to the inlet, and to a ring that is pinned to the oxidizer volute. A wear-ring on the volute mates against the impeller. The cavity between the volute and the impeller is called the balance cavity. Pressure in the balance cavity exerts a downward force against the fuel impeller to counterbalance the upward force of the oxidizer impeller, and so control the amount of shaft axial force applied to the No. 1 and No. 2 bearings. Leakage between the impeller inlet and the discharge is controlled by a wear-ring, which mates with the impeller and serves as an orifice. The fuel volute provides support for the bearing retainer, which supports the No. 1 and No. 2 bearings and houses the bearing heater. The No. 3 seal, which is set between the No. 1 bearing and the oxidizer intermediate seal, prevents the fuel for lubricating the bearings from coming into contact with the oxidizer. If fuel should pass the seal, purge flow from the oxidizer intermediate seal will expel the fuel overboard. On the fuel side of the No. 2 bearing, the No. 4 lube seal contains the lubricant within the bearing cavity. The remaining seal in the fuel volute is the primary seal, and it contains fuel under pressure in the balance cavity, maintains the desired balance cavity pressure, and prevents high pressure fuel from entering the low pressure side.

## Turbine

The 55,000 brake horsepower turbine drives the fuel and oxidizer pumps. It is a two-stage velocity-compounded unit consisting of two rotating impulse wheels separated by a set of stators. The turbine is mounted on the fuel pump end of the turbopump so that the elements of the turbopump possessing the greatest operating temperature extremes (1,500 degrees F for the turbine and −300 degrees F for the oxidizer pump) are separated. Hot gas from the gas generator enters the turbine through the inlet manifold at a rate of 170 pounds per second and is directed through the first stage nozzle to the 119-blade first stage wheel, then through the second stage wheel into the heat exchanger. This flow of hot gas rotates the turbine, which in turn rotates the propellant pumps. During mainstage operation the turbine speed is 5,550 rpm.

## Bearing coolant control valve

This valve incorporates three 40-micron filters, three spring loaded poppets and a restrictor. It performs two functions. Its primary function is to control the supply of coolant fuel to the turbopump bearings. Its secondary function is to provide a means to preserving the turbopump bearings between static firings of the engine or during engine storage. During engine firing, the coolant poppet opens and delivers filtered fuel to the turbopump bearing coolant jets, and the restrictor provides the proper turbopump bearing jet pressure.

4-10 An F-1 engine mockup photographed on January 18, 1966. It was identical to the contemporary production engines and was maintained on the assembly floor as a point of reference. (Rocketdyne, Harold C. Hall collection)

## GAS GENERATOR SYSTEM

The gas generator system provides the hot gases to drive the velocity-compounded turbine. It includes a gas generator valve, an injector, a combustion chamber and the propellant feed lines. The propellants are supplied to the gas generator by the No. 2 fuel and oxidizer outlet lines of the turbopump. Propellants enter the gas generator through the valve and injector, and are ignited in the combustion chamber by dual pyrotechnic igniters. The valve is hydraulically operated by fuel pressure from the hydraulic control system. Relative to the engine mixture ratio, the gas generator ratio is fuel-rich to provide a low combustion temperature in the uncooled gas generator and in the turbine.

### Gas generator valve

The hydraulically operated gas generator valve controls and sequences the entry of propellants into the gas generator. Hydraulic fuel is recirculated through a passage in the valve housing to maintain seal integrity and to prevent the fuel in the fuel ball housing freezing. Fuel is also recirculated through a passage in the piston between the opening port and the closing port in order to prevent the piston O-ring freezing.

### Gas generator injector

The injector directs fuel and oxidizer into the gas generator combustion chamber. It is a flat-faced multi-orificed injector incorporating a dome, a plate, a ring manifold, five oxidizer rings, five fuel rings, and a fuel disc. The gas generator valve and the gas generator injector fuel inlet housing tee are mounted on the injector. Fuel enters the injector through the gas generator fuel inlet housing tee from the gas generator valve. The fuel is directed through internal passages in the plate and injected into the combustion chamber by orifices in the fuel rings and the disc. Some of the orifices in the outer fuel ring also provide a cooling film of fuel for the combustion chamber wall. Oxidizer from the gas generator valve enters the injector through the oxidizer inlet manifold. It is directed from the oxidizer manifold through internal passages in the plate and is injected into the combustion by through the orifices in the oxidizer rings.

### Gas generator combustion chamber

The gas generator combustion chamber provides a space for burning propellants and directs the combustion gases into the turbopump turbine manifold. It is a single-wall chamber located between the gas generator injector and the turbopump outlet.

## ENGINE INTERFACE PANEL

Mounted above the LOX and fuel inlets of the turbopump, the engine interface panel provides the vehicle connect location for electrical connectors between the engine and the S-IC, as well as the attachment point for the flexible heat-resistant curtain. It is fabricated from heat-resistant stainless-steel castings made in three sections and assembled by rivets and bolts.

## ELECTRICAL SYSTEM

The electrical system includes flexible armored wiring harnesses for actuation of the engine controls and the flight instrumentation harnesses.

## HYDRAULIC CONTROL SYSTEM

The hydraulic control system operates the engine propellant valves during the start and cutoff sequences. It consists of a hypergol manifold, a checkout valve, an engine control valve and associated tubing and fittings.

4-11 Another view of the F-1 mockup viewing from the engine interface panel to the nozzle extension. White caps on the exterior of the thrust chamber protected the attachments for mounting the thermal insulation. (Rocketdyne, Harold C. Hall collection)

## Hypergol manifold

The hypergol manifold directs hypergolic fluid to the separate igniter fuel system in the thrust chamber injector. It includes the hypergol cartridge, an ignition monitor valve, a position switch and an igniter fuel valve. The cartridge, position switch and igniter fuel valve are all internal to the manifold. The igniter fuel valve is a spring loaded cracking check valve that opens to enable the fuel to enter, with the resulting pressure surge bursting the diaphragms of the cartridge and releasing the hypergolic fluid. A spring loaded cam-lock mechanism incorporated into the manifold actuates a position switch that indicates when the cartridge is installed and prevents actuation of the ignition monitor valve until after the diaphragms of the cartridge have burst.

## Ignition monitor valve

The ignition monitor valve is a pressure actuated, three-way valve mounted on the hypergol manifold. It controls the opening of the fuel valves, allowing them to open only after satisfactory combustion has been attained in the thrust chamber. When the cartridge is installed in the hypergol manifold, a cam-lock mechanism prevents the ignition monitor valve poppet from moving from the closed position. The ignition monitor valve has six ports: a control port, an inlet port, two outlet ports, a return port and an atmospheric reference port. The control port receives pressure from the thrust chamber fuel manifold. The inlet port receives hydraulic fuel pressure for opening the fuel valves. When the ignition monitor valve poppet is in the deactuated position, hydraulic fuel from the inlet port is stopped at the poppet seat. When the cartridge diaphragm bursts, the spring loaded cam-lock retracts to allow the ignition monitor valve poppet unrestricted motion. When thrust chamber pressure (directed to the control port from the thrust chamber fuel manifold) increases, the ignition monitor valve poppet moves to the open (actuated) position and hydraulic fuel is directed through the outlet ports to the fuel valves.

### Checkout valve

The checkout valve includes a ball, a poppet and an actuator. It provides for ground checkout of the ignition monitor valve and the fuel valves, and prevents the ground hydraulic return fuel, used during checkout, from entering the S-IC fuel tank via the engine system. In performing the engine checkout or servicing, the checkout valve ball is positioned so that fuel entering the engine hydraulic return inlet port will be directed though the ball and out the return port to the ground support equipment. For engine static firing, and in flight, the ball is positioned so fuel that enters the engine hydraulic return inlet port will be directed through the ball and out the engine return outlet port.

### Engine control valve

The engine control valve incorporates a filter manifold, a four-way solenoid valve and two swing check valves. The filter manifold has three filters: one in the supply system, and one each in the opening and closing pressure systems. They preclude entry of foreign matter into the four-way solenoid valve or the engine. Two swing check valves are plumbed into the supply system filter to permit hydraulic system operation using hydraulic fluid provided by the ground for checkout and servicing, or provided by the engine for normal engine operation.

The four-way solenoid valve is comprised of a main spool and sleeves to achieve two-directional control of the fluid flow to the main fuel, main oxidizer, and gas generator and valve actuators. The spool is pressure positioned by two three-way slave pilots, each of which has a solenoid controlled three-way primary pilot that is normally open. The de-energized position of the engine control valve facilitates hydraulic closing pressure to all engine propellant valves. Momentary application of 28 VDC to the start solenoid will initiate control valve actuations that culminate in the positioning of the main spool so that hydraulic pressure is applied to the opening port, and the pressure previously applied to the closing port is vented to the return port. A passage within the housing maintains a common pressure applied between the opening port and start solenoid poppet. After start solenoid de-energization, this

4-12 A view down the centerline of the F-1 engine mockup. (Rocketdyne, Harold C. Hall collection)

pressure holds the main spool in its actuated position, thus maintaining the pressure directed to the opening port without further application of the electrical signal to the start solenoid. Momentary application of 28 VDC to the stop solenoid will initiate control valve actuations that culminate in positioning the main spool so that pressure is vented from the opening port and applied to the closing port. The override piston may be actuated at any time by a remote pressure supply, which, in the event of an electrical power loss, would reposition the main spool and apply hydraulic pressure to the closing port. If electrical power and hydraulic power are both removed, the valve will return to the de-energized position by spring force. If hydraulic pressure is reapplied, pressure will be applied to the closing port. If an electrical signal is sent to the start and stop solenoids simultaneously, the stop solenoid will override the start and return the valve to a deactivated position.

### Swing check valve

There are two identical swing check valves on the engine control valve. They allow the use of ground hydraulic fuel pressure during engine starting transient and engine hydraulic fuel pressure during engine mainstage and shutdown. One check valve is in the engine hydraulic fuel supply inlet port and the other is in the ground hydraulic fuel supply inlet port.

## FLIGHT INSTRUMENTATION SYSTEM

The flight instrumentation system consists of pressure transducers, temperature transducers, position indicators, a flow measuring device, power distribution junction boxes and associated electrical harnesses, and permits monitoring of engine performance. It has primary and auxiliary elements: the primary is used for all static firings and vehicle launches; the auxiliary is used during research, development and acceptance portions of the engine static test program, and initial vehicle flights. The flight instrumentation system components, including both the primary and auxiliary systems, are listed below:

### Primary instrumentation

- Fuel turbopump inlet No. 1 pressure
- Fuel turbopump inlet No. 2 pressure
- Common hydraulic return pressure
- Oxidizer turbopump bearing jet pressure
- Combustion chamber pressure
- Gas generator chamber pressure
- Oxidizer pump bearing No. 1 temperature
- Oxidizer pump bearing No. 2 temperature
- Turbopump bearing temperature
- Turbopump inlet temperature
- Turbopump speed

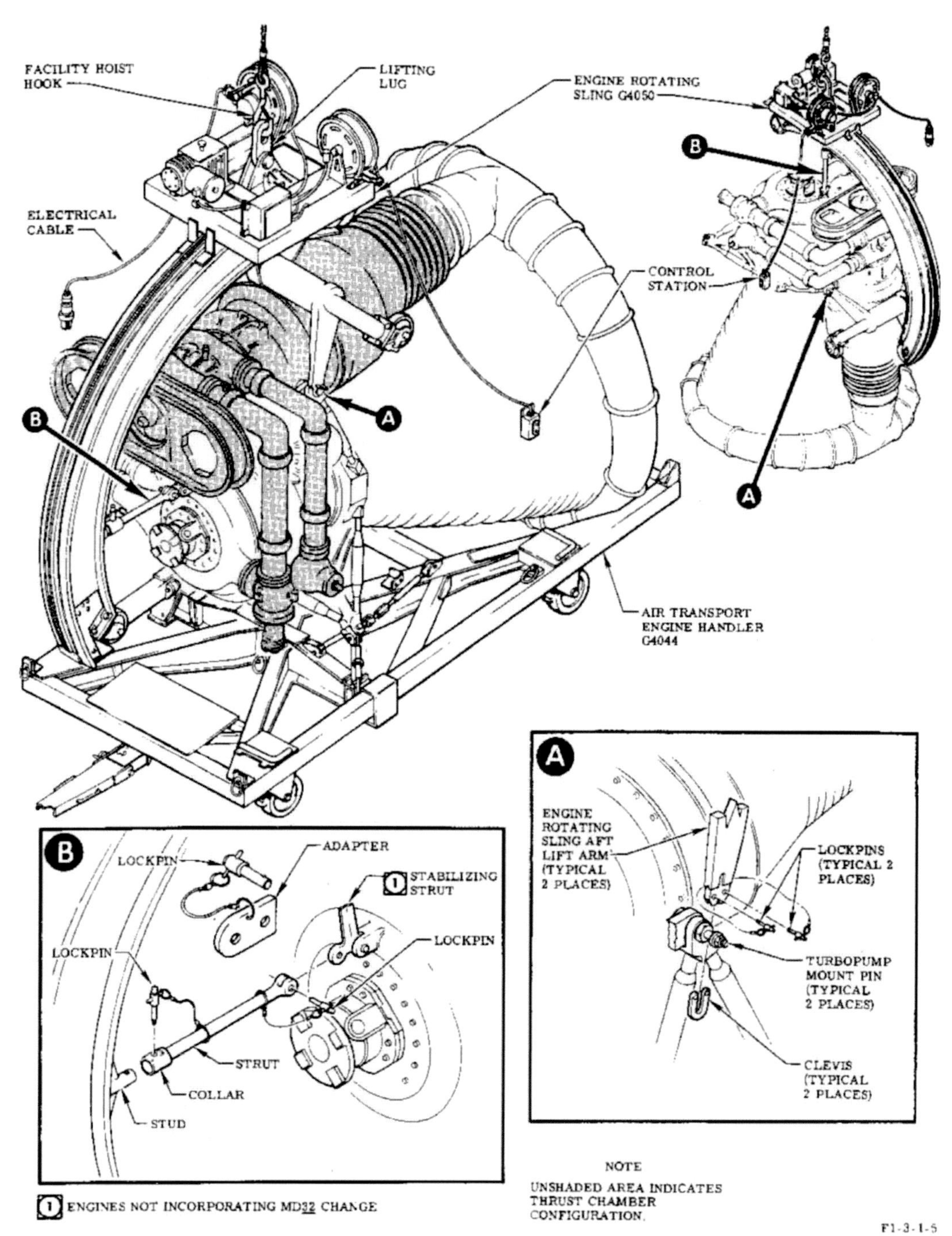


4-13 The Rotating Sling and Air Transport Engine Handler for the F-1 engine. (Rocketdyne, Vince Wheelock Collection)

### Auxiliary instrumentation

- Oxidizer turbopump seal cavity pressure
- Turbine outlet pressure
- Heat exchanger helium inlet pressure
- Head exchanger outlet pressure
- Oxidizer turbopump discharge No. 1 pressure
- Heat exchanger LOX inlet pressure
- Heat exchanger GOX outlet pressure
- Fuel turbopump discharge No. 1 pressure
- Engine control opening pressure
- Engine control closing pressure
- Heat exchanger LOX inlet temperature
- Heat exchanger GOX outlet temperature
- Heat exchanger helium outlet temperature
- Fuel pump inlet No. 2 temperature
- Heat exchanger LOX inlet flow rate

### Primary and auxiliary junction box

There are two electrical junction boxes in the flight instrumentation system. The primary has provisions for eight electrical connectors and the auxiliary for five. Both boxes are welded closed and pressurized with an inert gas to preclude admission of contaminants and moisture.

## ENGINE OPERATION

A ground hydraulic pressure source, thrust chamber prefill, gas generator, turbine exhaust igniters and hypergolic fluid are required to start the F-1 engine. To sustain its operation, it requires hydraulic pressure, electrical power and propellants.

When the start button is actuated, the checkout valve transfers the hydraulic fuel return from the ground line to the turbopump low pressure fuel inlet. The high level oxidizer flow is started to the gas generator and thrust chamber LOX dome. The gas generator and turbine exhaust gas igniters fire, and the start solenoid of the engine control valve is energized. Hydraulic pressure is directed to the opening port of the oxidizer valves. The oxidizer valves are partly open, and the hydraulic pressure is directed to the gas generator valve opening port. The gas generator valve opens, propellants under tank pressure enter the gas generator combustion chamber, and the propellant mixture is ignited by the gas generator igniters. The resulting exhaust gas is ducted through the turbopump turbine, the heat exchanger, and the thrust chamber exhaust manifold into the nozzle extension walls where the fuel-rich mixture is ignited by the turbine exhaust gas igniters. As the turbine accelerates the fuel and the oxidizer pumps, the pump discharge pressures increase and the propellants are supplied at increasing flow rates to the gas generator. Turbopump acceleration continues and, as

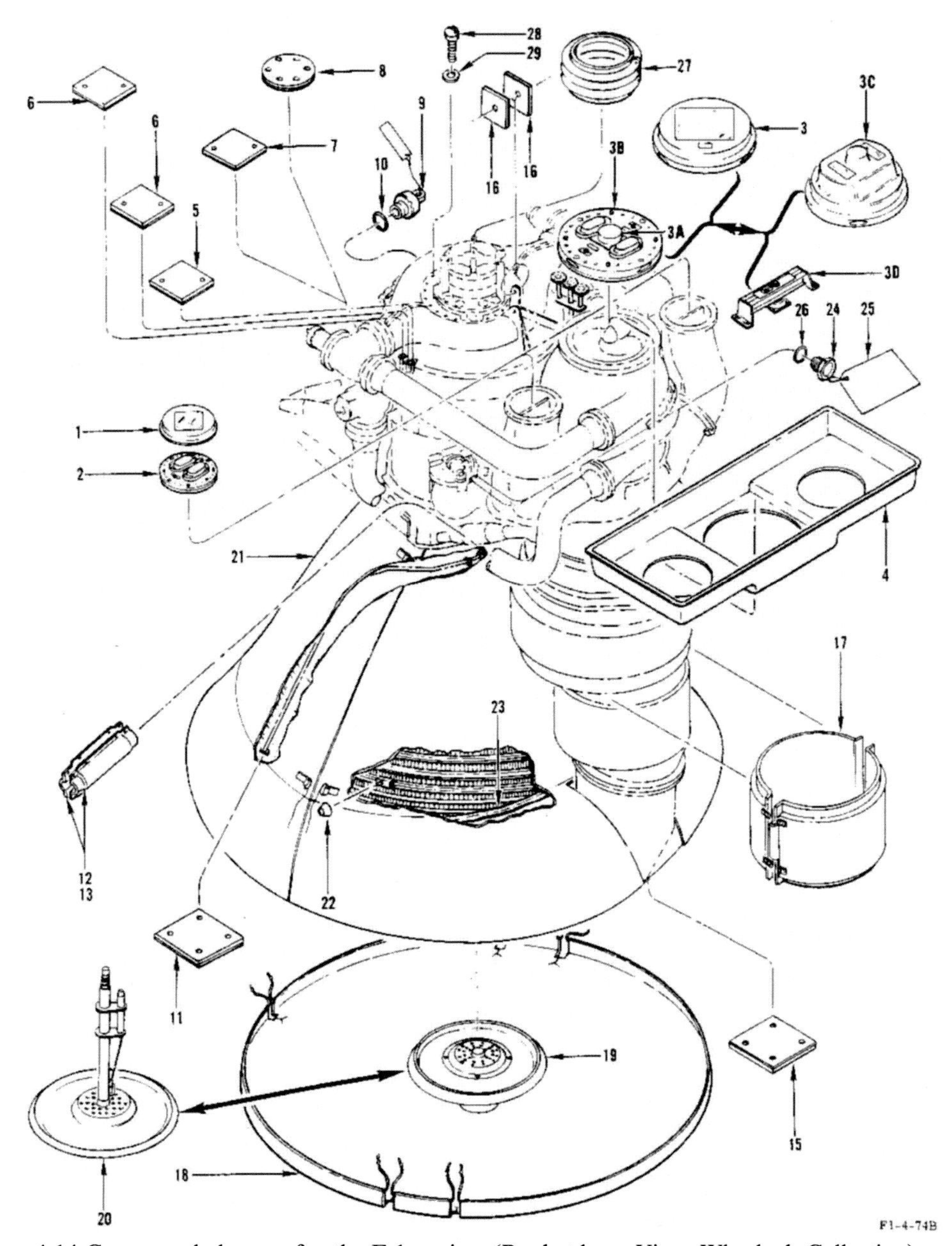


4-14 Covers and closures for the F-1 engine. (Rocketdyne, Vince Wheelock Collection)

the fuel pressure increases, the ignition fuel valve opens and allows fuel pressure to build up against the hypergol cartridge diaphragms, causing them to rupture and release the hypergolic fluid, which is fed into the thrust chamber together with the ignition fuel. When hypergolic fluid comes into contact with the oxidizer in the thrust chamber it spontaneous ignites to start fuel combustion. Thrust chamber pressure is transmitted through the sense line to the diaphragm of the ignition monitor valve, and when the thrust chamber pressure rises the ignition monitor valve actuates to allow hydraulic fluid to flow to the opening port of the fuel valves. The fuel valves open and fuel is admitted to the thrust chamber fuel inlet manifold. A proportion directly enters the injector to the thrust chamber, and the remainder passes through the tubing forming the wall of the chamber for cooling purposes on its way to the injector. As the thrust pressure increases, the thrust-OK pressure switches actuate to indicate the engine is operating satisfactorily. The thrust chamber pressure continues to increase until the gas generator reaches its rated power, controlled by orifices in the propellant lines feeding the gas generator. When engine fuel pressure exceeds the ground-supplied hydraulic pressure, the hydraulic pressure supply source is transferred to the engine. Hydraulic fuel is circulated through the engine components and returned through the engine control valve and checkout valve to the turbopump fuel inlet. When the fuel valves open, the ground hydraulic source facility shutoff valve moves to its closed position. This allows the engine hydraulic system to supply the hydraulic pressure during the cutoff sequence.

## ENGINE CUTOFF

When the cutoff signal is initiated, the LOX dome operational oxidizer purge comes on, and the stop solenoid of the engine control valve is energized. The hydraulic pressure that is holding open the gas generator valves, the oxidizer valves and the fuel valves is routed to return. Simultaneously, hydraulic pressure is directed to the closing ports of the gas generator valve, the oxidizer valves and the fuel valves. The checkout valve is actuated and, as propellant pressures decay, the high level oxidizer purge starts to flow. Then the igniter fuel valve and the ignition monitor valve close. The thrust chamber pressure will reach zero level at about the same time the oxidizer valves reach their fully closed position.

## ENGINE TRANSPORTATION

As each F-1 engine cost tens of millions of dollars, it had to be carefully handled, secured and protected in transportation. Rocketdyne designed support equipment for transporting engines by aircraft, by barge either on inland waterways or by sea, and by truck. The F-1 was transported by flatbed truck from the company's headquarters at Canoga Park in California to Edwards Air Force Base for testing, then airfreighted to MSFC or the Mississippi Test Facility as appropriate. From there it could be sent by barge to the Michoud Assembly Facility in Louisiana. The completed S-ICs were shipped by barge to the Kennedy Space Center. If replacement engines were needed at the Cape, they could be either flown or shipped by barge.

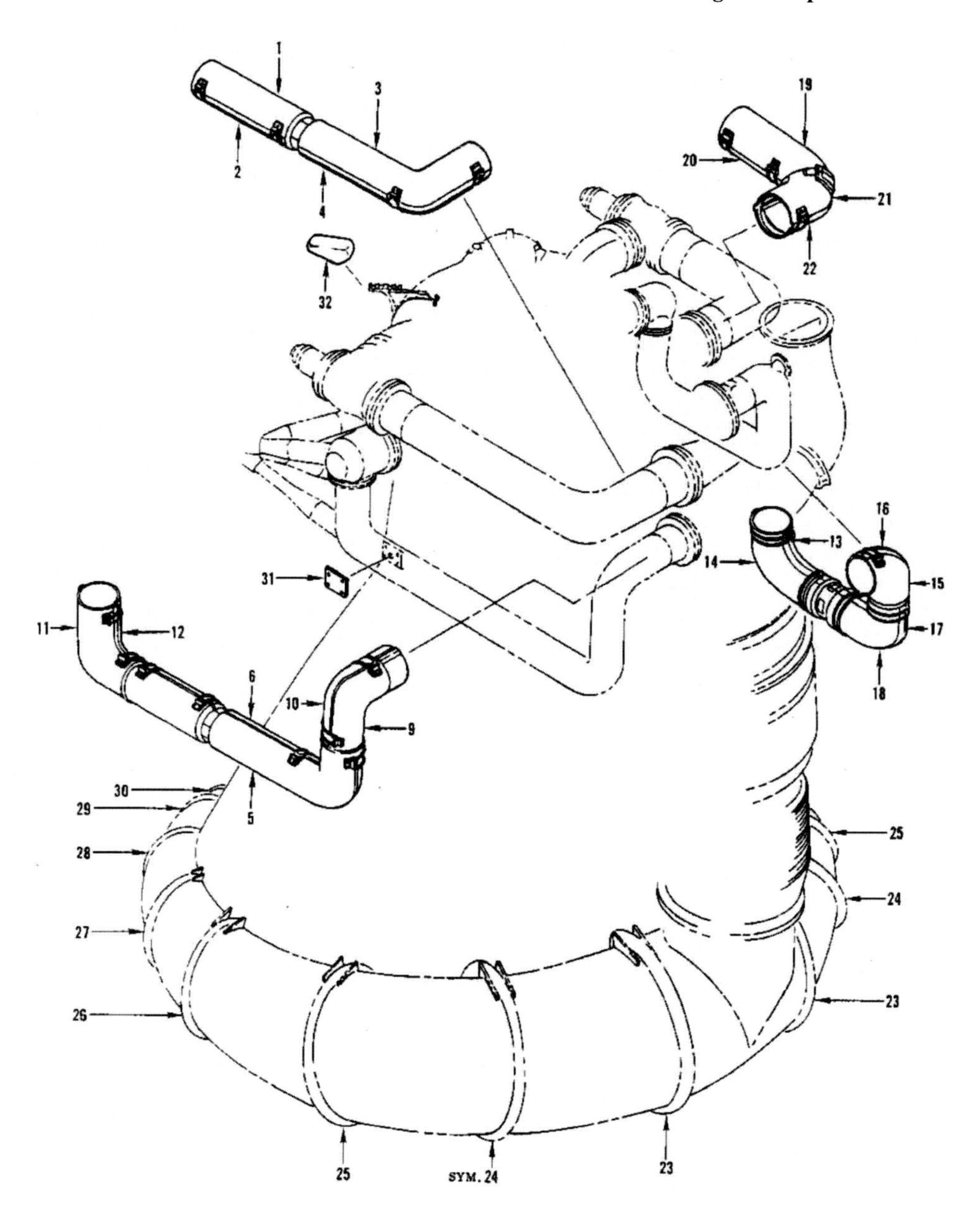


4-15 Thrust chamber manifold and propellant duct closures for the F-1 engine. (Rocketdyne, Vince Wheelock Collection)

## Engine handling support equipment

During the completion of F-1 engine assembly, protective covers and closures were secured to the engine. These were removed as the engine was installed on the S-IC. Gimbal Bearing Locks were installed to immobilize the gimbal bearing. The Thrust

Chamber Throat Security Closure was installed. In the Rocketdyne plant, the engine was handled and reoriented using the Engine Rotating Sling. This featured a semicircular track that permitted the secured engine to pivot from vertical into horizontal position for placement on the Air Transport Engine Handler (ATEH). A steel tube frame was then affixed to the ATEH and to the thrust chamber exhaust manifold with straps. Over this went the heavy fabric engine cover. The ATEH with its four steel wheels could be moved using a tow bar. An Engine Handler Sling was hooked to the four corners of the ATEH and lifted using a hoist for placement on a flatbed truck platform, where it was secured in place with chains.

The F-1 nozzle extension had its own dedicated equipment. The nozzle extension was first fitted with the Nozzle Extension Handling Fixture (NEHF), which provided internal structural support. The nozzle extension and the NEHF were then placed on Nozzle Extension Handling Adapter (NEHA). The Engine Handler Sling was then used to lift the NEHA for placement on the flatbed trailer, where it was secured with chains. Along with the engine and nozzle extension, various support equipment and the engine's thermal insulation were also shipped.

There was another vital item to protect the F-1 engine. During a checkout using the 500-F Saturn V at the Cape, a heavy rain storm drenched the thermal protective covers over the F-1 engines. If a Saturn V were to be launched with water-soaked thermal insulation, then the water would super-heat, rupture the thermal covers and expose the engine to dangerous temperatures. In order to protect the engines and their nozzle extensions and thermal insulation against inclement weather on the pad, Rocketdyne developed the F-1 Engine Environmental Cover, which consisted of a rubber-impregnated glass fabric cover with nylon drawstrings that had an aluminum coating on the inside. The cover had grommets at the top and bottom and two double rows down each side to permit the cover to be laced up around the engine. There were flaps to provide access to thrust chamber drain ports and several holes for drain lines.

## Engine transportation by aircraft

A number of specialized cargo aircraft were used to transport the F-1 engine and its nozzle extension to NASA facilities. At the airport, the F-1 engine on its ATEH and the nozzle extension on its NEHA were moved from the flatbed trailer to a dedicated shipping pallet and loaded onto a cargo lift trailer. For NASA's B-377-PG (Pregnant Guppy) and B-377-SG (Super Guppy), a tug positioned the cargo lift trailer in front of the opened aircraft, and raised it to the correct level, whereupon the pallet was moved into the aircraft and secured into position. F-1 engines were also transported aboard C-133 Cargomasters operated by the Air Force.

## Engine transportation by barge or ship

F-1 engines could also be shipped by one of the dedicated barges used by NASA to transport the various Saturn V stages or the Saturn I. A mobile crane or hoist was needed to remove the F-1 ATEH and NEHA from the truck flatbed for stowage on the enclosed barge.

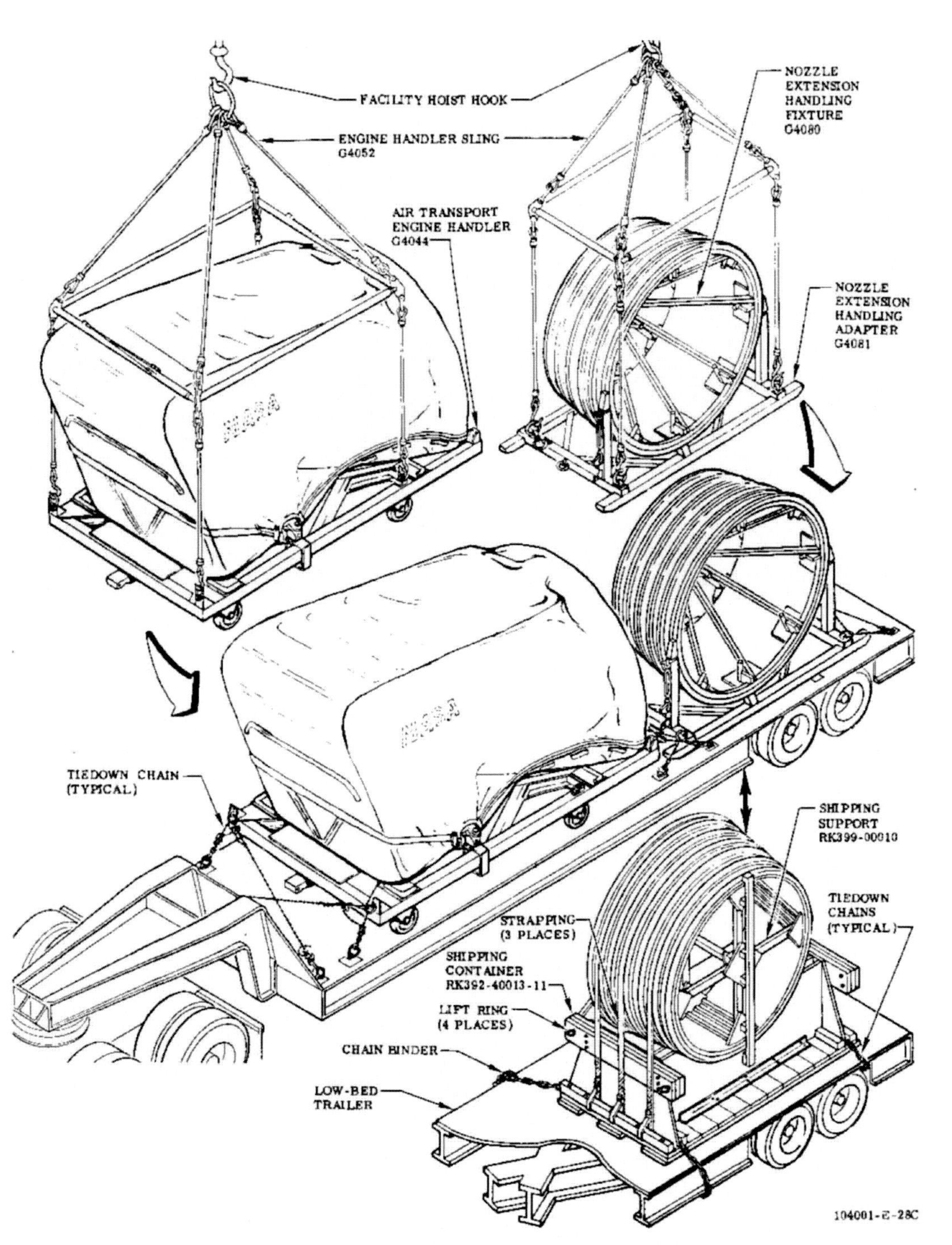


4-16 Loading an F-1 engine and nozzle extension onto a low-bed trailer prior to ground transportation. (Rocketdyne, Vince Wheelock Collection)

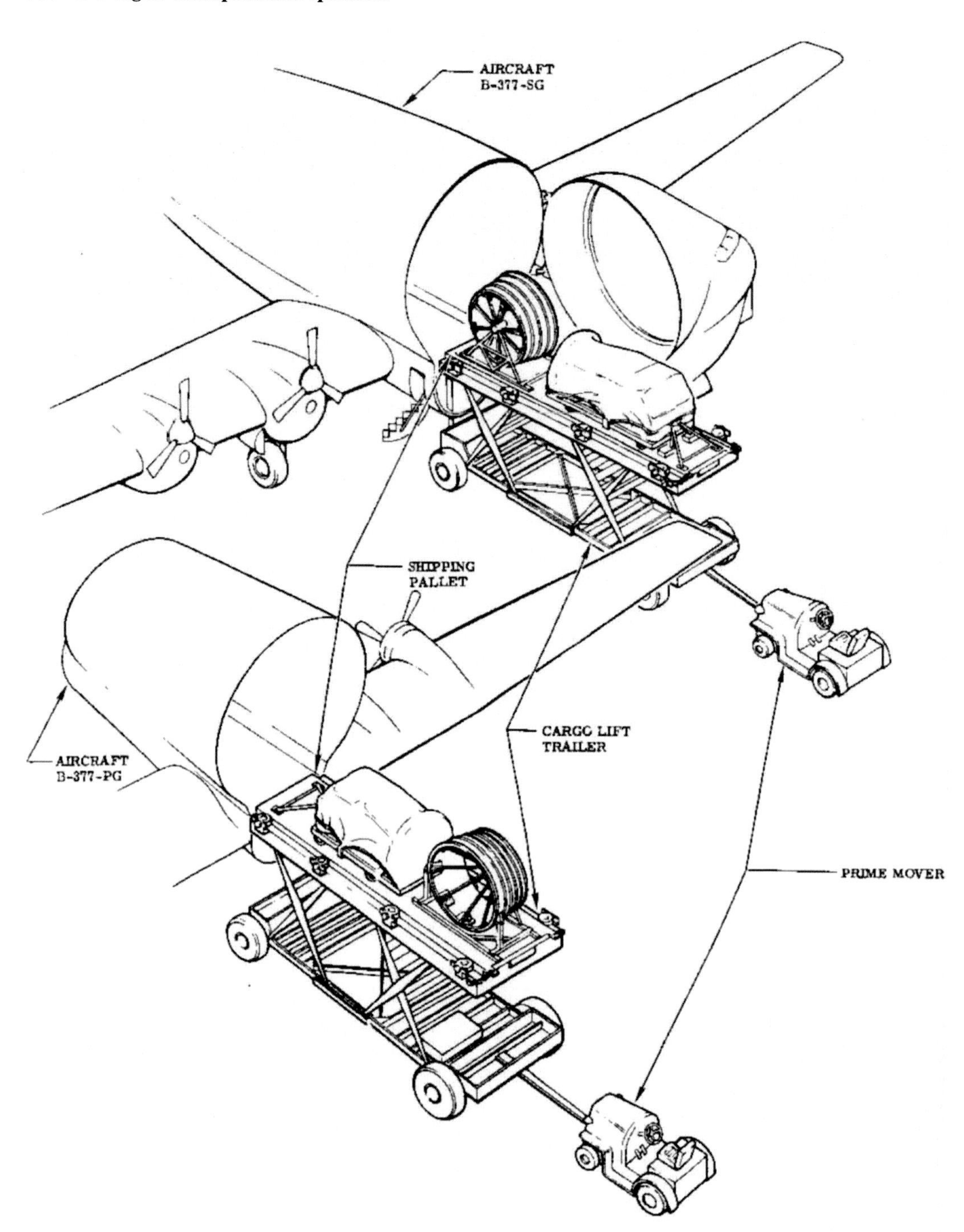


4-17 Positioning and loading an F-1 engine and nozzle extension shipping pallet aboard an aircraft. (Rocketdyne, Vince Wheelock Collection)

## Engine ground transportation by truck

NASA did not initially envision shipment of the F-1 engine and related equipment by truck. However, a plan to do so was development just in case one or both of the transport aircraft were lost to mishap, damaged or simply unavailable. A direct land route was the next fastest means of delivery. In fact, this plan was implemented, and engines were shipped directly from Canoga Park to NASA facilities in the central and eastern parts of the nation. In addition to the abovementioned preparations for shipment, for truck transport. The Turbopump Shaft Preload Fixture was installed on the turbopump oxidizer inlet. If a turbopump had to be shipped individually, the same fixture was installed.

# 5

# Manufacturing the F-1 engine at Rocketdyne

North American Aviation (NAA) was a prime contractor to NASA during the Apollo program. It was responsible for building the S-II stage of the Saturn V as well as the Apollo Command and Service Modules, and its Rocketdyne division manufactured the F-1 and J-2 rocket engines that, between them, powered all three stages of the launch vehicle. Rocketdyne also manufactured smaller rocket control and ullage (propellant settling) engines. As a result, NAA experienced phenomenal growth during the 1960s. In particular, employment at Rocketdyne's Canoga Park, California plant boomed. With the award of the initial F-1 production contract on July 2, 1962 for 55 engines, Rocketdyne had to expand its manufacturing and assembly facilities. The nature of the business required ongoing research and development of materials, manufacturing, inventory control, quality control and, of course, testing.

## ROCKETDYNE IN THE 1960S

In fulfilling government contracts for the Navaho, Atlas, Jupiter, Redstone, Thor and Saturn I (H-1) rocket engines Rocketdyne had built up a capable and efficient manufacturing capability, and this was brought to bear on the F-1 engine program, which involved a rapid pace, frequent design changes and usually short production runs. Often, even as production hardware was being manufactured and delivered, design changes resulting from ongoing development were being implemented. The organization of Manufacturing was structured specifically for such conditions. The Factory Manager headed the Manufacturing Team, and the superintendents of the Fabrication and Assembly and the Manufacturing Support line sections

5-1 Prior to furnace brazing, the F-1 was placed over a pressure bag designed to support the thrust chamber during the brazing procedure. (Rocketdyne, Vince Wheelock Collection)

reported to him. He was assisted in a staff capacity by Engine Program Representatives whose function was to ensure conformance to the plans and schedules of the company's various programs. Each Program Representative was dedicated to a specific engine program, and monitored it throughout its manufacturing phases, both experimental and production programs.

To expedite design and development programs, the manufacturing of production engines was separated from the experimental work in Fabrication and Assembly. The most experienced machinists in the experimental shop worked closely with the engineering groups, often making experimental hardware from simple sketches or rough mockups. The Sheetmetal and Processing Unit did metal fitting, welding and processing for experimental work. The Prototype Machine Shop took detailed dimensional drawings and created exact prototype hardware for fabrication by the production units, which used subsystem manufacturing operations. For example, one production unit would have primary responsibility for producing the complete turbopump—from the arrival of the raw aluminum castings to the final assembly and test of the component. This was facilitated by requiring all the equipment and machinery necessary to accomplish this to be physically located in that specific unit's manufacturing area. This saved time, improved production control, ensured that schedules were met and simplified the process of inventory management. The various components were brought together on a moving line in the Final Assembly Department. During final assembly, test cells that accommodated complete engines were utilized for high pressure checkout, with the turbopumps being calibrated and the flow through thrust chamber injectors checked.

Other departments at Rocketdyne were a vital link to the smooth flow of engine production. The Scheduling Department provided support to Manufacturing for pricing, the control of work hours and budgets. The Planning Department handled normal planning and production control operations. It also contained a specialized planning group which performed the Process Control function. Manufacturing Analysis prepared specialized manufacturing reports on new techniques and their application and serviceability, and in order to pinpoint and eliminate inefficiencies made time and motion, and tooling rate studies. A vital manufacturing department was Tooling Design, which fabricated both experimental and production tooling. It also supervised the procurement of outside vendors and services in support of basic tooling design. Since Rocketdyne could not perform 100 percent of the work to manufacture its rocket engines, it drew on the manufacturing Purchased Labor Unit to use the abilities of many outside suppliers and vendors in the greater Los Angeles area. The Tooling Department, which was separate from Tooling Design, assisted all of the company's departments. Its tooling engineers were experienced in solving manufacturing problems, and they worked together with Engineering to improve manufacturability. To ensure the performance and reliability of its rocket engines, Rocketdyne implemented procedures and operations which it referred to as Process Control. Although these were vital elements in the control of the entire manufacturing process, they were just part of the overall manufacturing strategy. Additional steps had to be considered when establishing product reliability and repeatability in manufacturing.

5-2 The F-1 thrust chamber was placed inside a cupola and then placed inside the furnace for the brazing procedure. (Rocketdyne, Vince Wheelock Collection)

## Mechanized production control

The first step in the complete quality control of a rocket engine was procedural. At Rocketdyne in the Apollo era, this was known as Mechanized Production Control (MPC). It was a comprehensively planned system composed of subsystems which integrated all phases of the plant operations. At its core was Mechanized Inventory Control (MIC), where information was made available on demand. Data collecting networks transmitted simple inputs to the magnetic disk file that updated company intelligence of any new requirements such as: the status and location of parts and tools, progress in procurement action, materials availability, status on completion of assemblies and details, parts quality, direct labor and accounting changes across the company. Continuously updated information was also used to produce weekly and monthly tabulated reports for distribution to supervisors and managers. The company upgraded its central computer systems several times during the 1960s as part of the ongoing process of improving the MPC process.

## Process control

Rocketdyne's way of controlling operations by documentation, drawings, sketches and specifications was known as Process Control. This was conceived and initially used to guide and control certain critical manufacturing operations, involving only about 10 percent of the parts. But then the company converted all its tooling setup sheets to assembly operation books. These served the shop areas as required, and were vital aids to the engineering drawings. Manufacturing process specifications were created to define the correct manufacturing description of the many different and complex engineering process specifications involved. These were used either alone or as interpretive instructions in the assembly operation books.

## The briefing program

In a design and development program as vital as the F-1 engine, revisions were to be expected, new materials would be introduced and new manufacturing processes incorporated. Every effort had to be made to keep all employees fully informed of such changes. In addition to training courses run by the Education Department at Rocketdyne, auxiliary training was provided which included the briefing program. This took the form of regularly scheduled meetings or courses using audio/visual materials on the entire range of manufacturing operations—including, to give just a few examples, the theory and operation of new sophisticated equipment, control methods, and department operation.

## The manufacturing development operation

The manufacturing development operation at Rocketdyne during the 1960s was a vital part of manufacturing process control. Composed of a balanced admixture of technicians and experienced shop personnel, the department which carried out this

5-3 Inspecting a brazed F-1 thrust chamber. (Rocketdyne, Vince Wheelock Collection)

operation performed a variety of tasks, including: research and developmental and support functions to bridge the gap between theory and practical applications in relation to upgrading conventional equipment; establishing operating parameters for new equipment; and the design or discovery of new equipment and processes. In the early 1960s, at least 40 such programs were successfully completed with the goal of improving the company's rate of achievement during the rest of the decade and into the 1970s. At all times, Rocketdyne used state of the art manufacturing technology, including electrical discharge machining, electrochemical machining and electron beam welding. It also pursued solid-state oscillators that were capable of carrying high peak currents at repetition rates exceeding those common in then-current electrical discharge machines. At pulse rates as great as 100 megacycles, rotating graphite electrodes operated in much the same fashion as end mills, giving better surface finish than was attainable by previous equipment. A high pulse rate used in combination with a high current density could vastly improve machining rates whilst upholding surface finish. The effort on the electrochemical machining process centered on improving both cutting rates and accuracy by means of much greater current densities and improved electrolytes. Additional improvements in this area centered on the development of sensors to pre-arcing, programmed short duration peak-reverse charge to deplete the electrode, and the adaptation of the electrochemical process to conventional equipment. In the area of electron beam welding, equipment was upgraded to improve weld repeatability, provide better tracking capability and reduce machine downtime.

### The machinability index

In 1956, just five percent of a rocket engine's total weight was comprised of exotic materials which required special machining or fabrication. However, by the end of the decade the proportion of such materials had increased to about 90 percent. The machinability index introduced by Rocketdyne to reflect the increases in materials strength and hardness, showed that conventional apparatus could barely cope with the super alloys and refractory materials that were being used in hardware for the F-1, J-2 and H-1 engines. To cope with this, the company installed more robust equipment having greater power and more accuracy, and replaced machining tools more frequently. In the case of the F-1 engine, the use of Inconel-X in the thrust chamber required the creation of new forming techniques, much harder tooling and special furnace techniques. The propellant injectors also required more refractory alloys—migrating from 4130 through 347 stainless steel and ultimately to Inconel, which required unconventional machining equipment and procedures. In fact, due to its size and power, the F-1 pushed the technological envelope in practically all areas of its manufacture.

## NON-DESTRUCTIVE TESTING

A cornerstone of quality assurance, reliability and performance of Rocketdyne's liquid propellant rocket engines was the use of non-destructive testing (NDT) of the materials in addition to the components that went into its engines. This was an essential phase of the manufacturing process, because a defective material or part could result in component or subassembly failure that could, in turn, result in the destruction of the engine. Each part had to pass rigorous testing and inspection. At Rocketdyne, NDT involved the Engineering Development Laboratory, the Quality Assurance Laboratory, and Production Inspection. The Engineering Development Laboratory (EDL) was responsible for design review specification and revision, failure analysis material evaluation, research and development into test methods, and the introduction of new test methods into production inspection. The Quality Assurance Laboratory (QAL) was responsible for NDT drawings, the training of inspectors, supporting production inspections, in-house and supplier surveillance, the support of field tests and the introduction of new test methods into production

5-4 A close view of the tube bundle and bands around an F-1 thrust chamber. (Rocketdyne, MSFC History Office)

inspection. The main function of the Production Inspection Department (PID) was to evaluate engine components, materials and processes in relation to engineering drawing requirements and specifications in support of manufacturing. Rocketdyne inspectors using advanced instruments and test equipment ensured both the rapid flow of hardware and reliable test evaluation. The PID also was responsible for the planning and sequencing of the test methods during fabrication.

Rocketdyne used a variety of NDT methods to evaluate materials and processes, encompassing both the entire electromagnetic spectrum and various other means to provide a qualitative or a quantitative evaluation of the material or process without altering the item being tested. Its engine reliability program began with design and manufacturing to ensure accurate, high strength hardware that would not fail while it was operating within the specified conditions. Extensive inspection to make sure that a part met its specified tolerances, finishes and manufacturing processes were integral steps. After individual components had been tested, an assembly could be subjected to extensive testing during development in an effort to discover inherent weaknesses. Successful development testing paved the way to a repeatable process for the production of an engine to acceptance testing. All of the NDT procedures developed for earlier engines were applied to the F-1, although in some cases they had to be modified to take account of the new engine's unprecedented size.

## Radiographic inspection

For the F-1 engine, radiography was used to detect internal defects in all of the Class-I weldments and high strength castings; to determine braze alloy distribution in the brazed thrust chamber and components; and to internally inspect electrical assemblies for missing or broken components. Inspection labs were located in each of the manufacturing buildings, as well as at the Santa Susana Field Laboratory. Parts and components were inspected in a specialized enclosure by placing film in a light-tight holder, positioning this closely beneath the item to be inspected, then beaming radiation from a radioactive isotope of Iridium-192 toward the target. The exposed film was processed and dried, and the negative image evaluated to pass or fail the inspected part or component.

## Liquid penetrant inspection

Rocketdyne used a number of liquid penetrant methods in performing inspections. The penetrant could either be a dye, or a fluorescent type consisting of oil-based or water-based constituents, and it could be applied by dipping, spraying or brushing. The water-based penetrants were used on parts that would come into contact with liquid oxygen or other active oxidizers. Before a part could be inspected, it had to be cleaned of scale, dirt, oil, paint or any impurity. Leak testing was accomplished by applying penetrant to the inner surface and a developer to the outer surface. All Class-I and Class-II weldments, tubing, castings and forgings required inspection in this manner, as did machined parts having a non-magnetic finish. Rocketdyne also devised a special high resolution fluorescent penetrant inspection procedure to reveal very fine cracks.

## Magnetic particle testing

This was used to detect discontinuities in ferromagnetic materials. The procedure involved establishing a suitable magnetic field in the test item, applying magnetic particles to its surface, and inspecting for accumulations of particles. This method could also reveal discontinuities beneath the surface. The magnetic field could be

5-5 This photograph dated December 7, 1961 shows thrust chambers which have been through the furnace braze operation. (Rocketdyne, Harold C. Hall Collection)

oriented longitudinally or circumferentially. After inspection, a part had to be demagnetized prior to further processing.

## Ultrasonic testing

Ultrasonic inspection was performed by either transmission or pulse-echo. Testing was performed at frequencies between 1 and 25 megacycles per second. Ultrasonic testing and inspection was used to detect flaws in thin or thick plates, bars, rods, forgings, tubing and weldments. Thicknesses ranging from less than a millimeter to several meters could be ultrasonically tested, but most tests were on thicknesses up to 15 centimeters in accordance with ASTM E-113. This means of inspection had the advantages of:

1. High sensitivity, permitting detection of minuscule defects.
2. Great penetrating power.
3. Accuracy in the measurement of flaw position and estimation of flaw size.
4. Fast response, permitting rapid and automated inspection.
5. Need for access to only one surface of the specimen.

The application of ultrasonic testing for inspection of weldments and brazed assemblies grew considerably at Rocketdyne during the F-1 engine program. Some welded assemblies were not inspected by radiography owing to their thickness or the geometry of the part, and were inspected ultrasonically instead. This method was used extensively to detect defects in brazed injectors, exhaust gas generators, stators and thrust chamber tube-to-jacket and tube-to-band braze joints. It was used with all forgings, Rene 41, Inconel-X and Hastelloy-C plate stock. All Inconel-X thin-wall seamless thrust chamber tubing was inspected for longitudinal defects. Rocketdyne built an automated system for inspecting straight and straight-tapered tubing.

### Eddy current testing

In eddy current testing, an alternating magnetic field induced eddy currents in the test object, in turn making an electromagnetic field in the immediate vicinity of the test object. The range of physical properties that could be measured included alloy variation, heat treatment, hardness, the magnitude of defects, dimensional changes, conductivity and permeability. It was effective in sorting out different alloys, or a single material of different tempers. It was helpful in testing thin, nonferrous metal tubing or sheets.

### Infrared testing

Infrared testing was done on selected components of the F-1 engine. Every object radiates heat at some range of wavelengths. When a component was altered by the manufacturing process, or there was a defect present, the heat that it emitted would vary across its surface. Heating a part and then scanning it for variations during the cooling process could reveal such flaws. Infrared testing could also be done while the part or component was in the process of being heated. Because Rocketdyne did not have the equipment to do these tests, the task was contracted to a company in Boulder, Colorado.

### Kinefluorography and cinefluorography

These methods of inspection beamed X-rays through the test object, and observed it with a fluoroscopic image intensifier. Electronic components and thermocouples were inspected using kinefluorography. Cinefluorography using motion picture cameras were used at Rocketdyne to evaluate failure modes in small ablative thrust chambers during hot firing tests.

### Thickness testing devices

Rocketdyne used several different instruments to measure the thickness of plating or materials: including the Permascope, which measured nonmagnetic coatings on a ferric substrate; the Dermitron, which operated by setting up an eddy current and measured either anodize on aluminum or nonconductive coatings on nonmagnetic substrates; the Betascope, which measured the thickness of a coating of either gold on copper or chromium on aluminum; and the Process Nucleonics Thickness Gage that

5-6 The F-1 thrust chamber underwent many non-destructive testing inspections during engine assembly. Here, technicians are ultrasonically inspecting tube-to-shell brazing bonds. (Rocketdyne, Vince Wheelock Collection)

utilized the gamma ray backscatter principle to measure the thickness of small diameter, tapered, thin-wall thrust chamber tubing.

### Hardness testing

Rocketdyne used several different machines to check hardness of heat-treated parts to ensure that they conformed to the specifications on the drawing. In addition to the industry standard Rockewell hardness tester, Rocketdyne employed Brinell and Riehle testers. It had a sizable metallurgical laboratory to test the surface finishes of specific F-1 engine components.

### Visual inspection

Visual inspection has always been a key aspect of quality control. At Rocketdyne, visual inspection of machined parts and manufactured components was aided by optical magnifiers, boreoscopes and fiber-optic scopes. Inspected components had first to pass a visual inspection prior to being subjected to further testing.

### Selection of the test method

The method of non-destructive testing to be used for a given part or component was stated on the drawing or other documentation, based on Rocketydne's years of manufacturing experience. All forgings required ultrasonic and penetrant testing or magnetic particle inspection. Class-I weldments needed radiographic or ultrasonic and penetrant or magnetic particle inspection. If a new part could not be tested by established NDT means, then the design engineer, materials and process engineers and Manufacturing and Quality Control departments had to coordinate to develop a new means of testing. Research and Development hardware naturally underwent the same NDT tests as were to be applied to the production hardware. The completed development components or assemblies were statically and dynamically tested to ensure conformance to design specifications, with failed components undergoing a failure analysis which included metallurgical evaluation. Inspection of production hardware required the consistent application of NDT methods every time for every part and component.

### F-1 engine reliability and NDT

The non-destructive testing of essentially every manufactured component that was integrated with other components into subassemblies for the F-1 was crucial to the engine's reliability and performance—as indeed it was for every rocket engine that the company made. The failure of a single part could trigger a chain of events that ended in the automatic shutdown of the engine or its destruction, in either case jeopardizing the mission. Rocketdyne strove to avoid the catastrophic failure of a crucial component by using the most stringent NDT. This, coupled with rigorous development testing, and then by production engine acceptance tests, resulted in the flight F-1 engines having a 100 percent success rate.

## THE F-1 THRUST CHAMBER

Rocketdyne's early studies of the F-1 engine established that it would present no insurmountable problems in either its manufacture or its ability to generate a thrust in excess of one million pounds. But its size did present manufacturing challenges unlike any the company had faced with its previous engines. The F-1 was the first engine to use extremely hard Inconel-X in the thrust chamber components, posing difficulties in the fabrication and brazing processes of the tube bundles. Combined with this, was the unprecedented size of the thrust chamber. Early tubular-wall regeneratively cooled rocket engines had used high heat transfer pure nickel tube bundles which were brazed by hand using low melting silver-based brazing alloys.

These conventional materials and processes could not be applied to the F-1 thrust chamber.

## The furnace brazing decision

Initial development work on joining processes established that conventional torch brazing techniques used on pure nickel would not work on Inconel X-750 owing to the characteristic low ductility of high-nickel alloys in the temperature range of most silver-base brazing alloys—i.e. 1,200 to 1,500 degrees F. Brazing specimens with Inconel X tubes and conventional brazing alloys revealed liquid metal stress cracking using low melting silver-based brazing alloy. After many specimen tests and analyses, it was realized that the only viable method of joining the tube bundle would be furnace brazing with high temperature brazing alloys. But the aluminum and titanium in Inconel X-750 tended to form refractory oxides which made the surface resistant to effective wetting by most brazing alloys. The metallurgists at Rocketdyne eliminated this issue by a means of electrolytically depositing a layer of nickel that was 0.0010 to 0.0014 inch in thickness on the surface of the Inconel X tubes and other parts.

The furnace built by Rocketdyne for brazing the F-1 thrust chamber was the largest of its kind. The decision to use furnace brazing gave additional advantages over traditional torch brazing by a human operator:

1. Minimizing thermal stresses resulting from differences in heating and expansion rates.
2. Uniform temperature applied over the entire surface of the thrust chamber during the brazing process providing consistency for each thrust chamber.
3. The opportunity to combine the brazing process with the age hardening of Inconel X-750.

## Selection of a thrust chamber brazing alloy system

Rocketdyne's decision to use Inconel X-750 for the thrust chamber, and furnace brazing as the primary method of joining the thrust chamber components, led to careful evaluation of potential alloy systems. Among those considered were silver-based, nickel-based and gold-based brazing alloy systems. Analysis of the thrust chamber brazed joint reliability requirements led to the decision to use a three-step brazing process. In the first brazing operation, the pair of smaller thrust chamber tubes would be joined to the larger primary tube by using induction heating. This subassembly would then be stacked and furnace brazed as part of the overall thrust chamber assembly. In the second brazing operation, all the thrust chamber parts, including the tube subassemblies, jacket, bands and engine rings, would be joined. In the third brazing operation, the thrust chamber would be partially re-alloyed in certain vital joint areas and then subjected to another furnace brazing operation. At each stage, the temperatures had to be carefully controlled in order not to melt the previously brazed components.

5-7 Fabrication and installation of the turbine exhaust manifold at the base of the F-1 thrust chamber. (Rocketdyne, Vince Wheelock Collection)

## Detail part preparation

All of the thrust chamber components, including the jacket, injector ring and exit ring, were precision machined. The forged and welded construction of the Inconel X-750 jacket was subjected to rigorous process requirements and quality control during fabrication. The thrust chamber tubes of Inconel X-750 were made to tight tolerances, and then induction brazed using a semi-automatic process. The bands which surrounded the thrust chamber nozzle extension were also precision formed. This precision in manufacturing was essential to ensuring that the entire chamber assembly was capable of satisfying the minimum clearances necessary to establish capillary joints and good brazing alloy flow. To achieve this, it was necessary that all parts received proper surface preparation. In order to ensure successful brazing of the Inconel X-750, the thrust chamber jacket and tube bundle were nickel plated by the electrolytic process. All thrust chamber components, included those which had been plated, were subjected to a succession of cleaning operations involving degreasing, alkaline cleaning, and a final rinse using deionized water to optimize brazing conditions. The thrust chamber components were then ready for assembly. In order not to jeopardize the strict requirements for successful brazing, the thrust chamber was assembled and alloyed for furnace brazing in a 'white room' that was used only for this purpose. The assembly of the thrust chamber tube subassemblies into the jacket and end rings, referred to as the 'stacking operation,' was done in a stacking fixture—a precision tool that established the initial alignment between the tube bundle, jacket and end rings prior to the brazing alloy application. The high precision in manufacturing and assembly prior to brazing is apparent from the fact that there was about 3,000 feet of tube-to-tube joint length to be sealed against the gases of the combustion chamber, and there were about 7,000 tube-to-band joints to be bonded. The gases were at a temperature of about 5,000 degrees F, and containing them in the combustion chamber was a significant accomplishment for the brazing process. For the first cycle brazing operation, the brazing alloy was in powdered form and was applied by a proprietary Rocketdyne spraying process. All phases of the thrust chamber furnace brazing process were controlled by the company's F-1 Furnace Brazing Process Specification. In the spraying process, alloy was applied to the tube-to-tube joints on the interior and exterior of the thrust chamber. The chamber was then ready for the furnace braze tooling setup.

## The brazing retort and high temperature pressure bag tooling

The unprecedented size of the F-1 thrust chamber had a primary requirement in the ability of the tooling to support the interior of the thrust chamber tube stack at temperatures in excess of 2,000 degrees F for successful brazing. This was done using a device called a 'pressure bag', which consisted of a flexible heat-resistant alloy skin, shaped to the interior contour of the thrust chamber, which, when it was inserted and pressurized, supported the tube stack. This technique was a significant departure from earlier concepts, which had employed a high mass rigid mandrel of heat resistant alloy. The good internal support characteristics and low mass of the pressure bag tooling were the primary factors in establishing a relatively rapid and economically feasible furnace brazing cycle. The brazing retort fulfilled a second requirement in the furnace brazing operation: the elimination of oxygen from the atmosphere surrounding the brazed assembly. This required the containment of an extremely pure protective atmosphere around the thrust chamber. The retort was constructed of a high nickel-chromium oxidation resistant stainless steel developed for this purpose. Its form was designed to maintain a protective atmosphere around and within the thrust chamber throughout the brazing cycle. Prior to initiation of the heating cycle, a gas line fed argon gas into the bottom of the retort to displace any residual air.

### Furnace brazing: the first cycle

The furnace brazing operation represented the final step in which all the hundreds of parts and subassemblies became a complete thrust chamber. In many respects it was similar to the launch of the Saturn V itself, as the failure of any of the many numerous control exercised during the furnace brazing operation could produce an unacceptable result. Consequently, the entire process was a highly developed and closely controlled operation. The furnace designed for the F-1 thrust chamber was unique in its design and performance. Its low mass construction provided a design with minimum heat capacity, and rapid response to the heating requirements of the thrust chamber during the brazing cycle. Its ability to provide uniform heating and rapid response was aided by that fact that it had three zones of temperature control. These performance characteristics made it possible to initiate the furnace brazing cycle with the furnace at room temperature, thereby eliminating the thermal shock that the thrust chamber would otherwise have been subjected to by being put into a furnace at its operating temperature. This cold-start capability helped to limit the gradients of the thrust chamber to 150 degrees F as the temperature was raised for brazing. Since thrust chamber component wall thicknesses varied from as little as 0.020 inch in the tube wall, to as great as 0.500 inch in portions of the jacket, large temperature gradients in a conventional furnace could have produced unacceptable thermal stresses and dimensional variations of parts. Following the first brazing cycle in the furnace, the thrust chamber was prepared for the first tube-to-tube leak test. Incomplete joints discovered during these tests or by radiographic inspection, were realloyed prior to the second cycle furnace brazing.

### Cleaning and alloy application for the second furnace brazing cycle

After the first brazing cycle, the thrust chamber was thoroughly cleaned and rinsed to remove any post-brazing residue and loose alloy particles, then returned to the white room in order for selected areas to be realloyed. The second cycle alloying process was accomplished using a slurry of gold-18 nickel alloy with a paste flux and alcohol. To optimize the flow and sealing capability of the brazing alloy, the thrust chamber was inverted with the jacket down for the second furnace brazing cycle. The second cycle differed from the first in that: because the thrust chamber was now self-supporting there was no requirement for internal support tooling; to preclude remelting the

5-8 A turbine exhaust manifold undergoes inspection in the optical tooling dock in November 1962. (Rocketdyne, Vince Wheelock Collection)

first braze, the brazing temperature was 1,800 degrees F, which was considerably lower than the first brazing cycle; and the Inconel X-750 had aged during the cooling cycle.

## Furnace brazing instrumentation

Thrust chamber temperature measurement was accomplished with chromel-alumel thermocouples, with calibrated checkpoints throughout the chamber to monitor its temperature. Thermocouples were also emplaced at each point where the thickness of a section, or its proximity to the furnace wall, might cause a variation from the average chamber temperature. To maintain the required highly pure atmosphere in the brazing retort, the thermocouple cables were fed through a line from the retort to a cool location at the base of the furnace, and then through a hermetic seal to the recording instruments. Because the most significant cause of surface oxidation and poor brazing was likely to be the presence of water vapor, the argon for furnace brazing the F-1 thrust chamber had less than four parts per million of water vapor. Any moisture entering or leaving the retort during furnace brazing was monitored by an electrolytic hygrometer, which measured the dew point of the argon in the retort. A cold cup containing dry ice and acetone was used as a low-tech backup to this instrumentation. The gas in the retort was also monitored for oxygen content and specific gravity.

After the second cycle in the furnace brazing, the thrust chamber was subjected to a series of pressure, penetrant and radiographic tests to establish that its brazed joints did not allow any leaks of hot gas and fuel. Also, pressure tests for hot-gas leakage in the tube-to-tube joints were made, radiographic inspections of the tube joints in critical areas of the thrust chamber were made, followed by penetrant tests of joints for fuel leakage and a static hydraulic pressure test for fuel leakage.

## THE F-1 MARK 10 TURBOPUMP

More design and fabrication time was spent on the Mark 10 turbopump of the F-1 engine than any other of its components. Simplicity and reliability were achieved by designing a direct drive unit with a single two-stage turbine driving both the fuel and oxidizer pumps. The material for the manifolds and rotors was selected for its high strength at elevated temperatures and mechanical properties in heavy sections. The turbine blade material was selected for its high strength at elevated temperatures, good mechanical and thermal fatigue resistance, and for being castable. The pump shaft material had to be readily fabricated, and have high strength and toughness at the operating temperatures. Oxidizer system components such as the inducer and impeller had to provide high strength at cryogenic temperatures. Weight reduction was always a prime consideration in evaluating materials. A number of the F-1 engine's cast aluminum components were cast by various foundries that had done such work for Rocketdyne over many years. Some of these cast components were delivered to Rocketdyne for machining and other manufacturing steps, but other aluminum components were delivered as finished parts. All such parts underwent inspection, and had to pass non-destructive testing procedures prior to assembly. All turbopump rotor components were either balanced as individual components prior to turbopump assembly, or thereafter by rotation in a spin pit. After assembly at Canoga Park, each turbopump was taken to the Santa Susana Field Laboratory for hot fire testing and then returned to Canoga Park for installation on an engine. This turbopump hot fire (and calibration if required) gave assurance that this key functional component would operate to specification during engine acceptance test hot fires. The oxidizer dome, propellant valve bodies and related components were also cast by outside foundries. Typically these were delivered to Rocketdyne in a rough cast or semi-finished state, and finish-machined to be certain of controlling all the vital precise machining steps to match the manufacturing drawings. The all-important injector was fully machined and finished at Rocketdyne, as was the F-1 engine gimbal assembly.

## HEAT EXCHANGER AND TURBINE EXHAUST MANIFOLD

The heat exchanger and turbine exhaust manifold for the F-1 engine were made using high strength refractory alloy steel in Rocketdyne's own sheetmetal shops. Individually formed sheetmetal components were precision welded to produce the finished components. The task was complicated by the intricate compound curves, in particular on the turbine exhaust manifold, which had to be carefully produced using special die stamps or special sheetmetal roller equipment. The finish-formed parts were checked against templates prior to being mated in fixtures for welding. The assembled heat exchanger and turbine exhaust manifold were subjected to thorough inspections and non-destructive testing to check for component integrity.

## HYDRAULIC LINES, HOSES AND WIRING HARNESSES

Rigid hydraulic lines were fabricated at Rocketdyne by precisely forming special stainless steel alloy tubing to pattern. Hydraulic or pneumatic hoses were produced by outside vendors that specialized in making such items for aerospace companies. Wiring or cable harnesses were generally fabricated by Rocketdyne, using pattern templates to assure correct cable and harness segment lengths, and terminated with Mil-Spec grade connectors. After all the connections were soldered, the connectors were often potted to ensure that they did not become contaminated and to prevent shorts. All finished cable assemblies or wiring harnesses were inspected and tested before being installed on the engine. One of the final engine assembly steps was to fit the protective covers, which, with other non-flight hardware, were individually designated and generally red in color to assist operations personnel in identifying items to be removed before assembly, test and flight.

## F-1 ENGINE FLOW

Rocketdyne published seven technical manuals as field support documentation for the description, operation and maintenance of the F-1 engine. The first, R-3896-1, contained all the information for the individual engine system components, engine operation, mounting on the S-IC stage of the Saturn V, the F-1 engine flow from NASA acceptance tests at Rocketdyne right through to delivery of the S-IC to the Kennedy Space Center, and other information related to the F-1 engine. Below is the *verbatim* text on F-1 Engine Flow from R-3896-1.

5-9 An assembled F-1 turbopump is lowered onto a dolly for transport to the next engine assembly area. (Rocketdyne, Vince Wheelock Collection)

Section 1-135.

The following describes F-1 engine flow and events that take place from the time of Customer acceptance of the engine at Rocketdyne, Canoga Park, through Apollo/Saturn V launch at Kennedy Space Center (KSC). After official acceptance of the engine (signing of DD Form 250), modifications may be made or maintenance tasks may be performed, with Customer approval, before shipment. The engine, nozzle extension, and loose equipment are shipped to the Michoud Assembly Facility (MAF) by either truck or ship. (Thermal insulation (TIS) is shipped to MAF by truck.) At MAF the engine is inspected and then assigned to a stage, designated as a spare, or left unassigned. Spare engines and unassigned engines are processed to a specific condition and placed in storage until needed. The normal flow of assigned engines consists of installing loose equipment and TIS brackets, performing modifications and maintenance, and installing the thrust vector control system on outboard engines. Single-engine checkout is performed, wrap-around ducts and hoses are installed, and the engines are installed in the stage. The stage and nozzle extensions are then shipped to the Mississippi Test Facility (MTF) by barge.

Section 1-136.

The stage is installed in the static test stand at MTF where the engines are inspected, and the nozzle extensions, slave hardware, and static test instrumentation are installed. A pre-static checkout of the stage is performed, followed by a static test, to determine stage acceptability and flight readiness. After a successful stage static test, the engines are inspected, test data is reviewed, and the turbopumps preserved. The nozzle extensions, slave hardware, and static test instrumentation are removed; then the stage is removed from the test stand, and the stage and nozzle extensions are shipped to MAF by barge. During normal stage flow at MAF, the installed-engines are inspected and refurbished; then a post-static checkout and a pre-shipment (to KSC) inspection are performed. The stage may be stored at MAF after engine refurbishment, depending on the stage schedule. The stage, nozzle extensions, loose equipment, and TIS are shipped to KSC by barge.

Section 1-137.

At KSC the stage is erected onto the Launch Umbilical Tower (LUT) in the Vertical Assembly Building (VAB), where a visual inspection is performed.[1] Loose equipment is installed, modifications are made, and maintenance tasks are performed. Stage and engine leak and functional tests are performed, and final installation of the TIS is completed. While the first stage is being prepared, other

[1] Author's note: The stage was actually erected on the Mobile Launch Platform, inside the Vehicle Assembly Building (the initial term Vertical Assembly Building for this facility was superseded; although there *was* a Vertical Assembly Building for the S-IC at the Michoud Assembly Facility). In what follows these points will not be corrected.

tasks are being done to prepare the remaining stages and modules, and the spacecraft, to mate and assemble them into the complete Apollo/Saturn V Vehicle. The vehicle and mobile launcher are then moved from the VAB to the launch pad on the crawler transporter, where launch preparations and final checkouts are performed. With all preparations complete and all systems ready, the Apollo/Saturn V is launched. After launch, a post-flight data evaluation is made, to determine that the S-IC stage engines operated within the specified values during vehicle launch.

1-138. ENGINE FLOW BEFORE FIELD DELIVERY.

1-139. Customer acceptance inspection.

1-140. Customer acceptance inspection is performed when Contractor engine activity at Canoga Park is complete. The Customer reviews all documentation including Component Test Records, Engine Buildup Records, Engine Test Records, and Engine Acceptance Test Records in the Engine Log Book. The Customer verifies that the engine configuration information on the engine MD identification plate corresponds to that listed in the Engine Log Book, and upon acceptance of all records and documentation, signs DD form 250, which constitutes official acceptance of the engine by the Customer.

1-141. POST-DD250 MAINTENANCE OR MODIFICATION.

1-142. If required before field delivery of an engine, post-DD250 maintenance or modification, as required by Engineering Change Proposals and Engine Field Inspection Requests can be done at Rocketdyne with Customer approval. Upon completion of maintenance or modification, the Engine Log Book is updated and the engine is accepted by the Customer.

1-143. ENGINE SHIPMENT TO MAF.

1-144. The engine, nozzle extension, and loose equipment is shipped to MAF by truck or ship as directed by the Customer. Detailed requirements for shipping the engine are in R-3896-9. Detailed requirements describing the use of handling equipment are in R-3896-3.

1-145. PREPARATION FOR SHIPMENT.

1-146. Preparation for shipment at the Contractor's facility consists primarily of removing the engine from buildup and test equipment, installing the engine and nozzle extension in shipping equipment, and packaging the loose equipment. Engine Rotating Sling G4050 is installed on the engine and a facility hoist lifts the sling and rotates the engine from vertical to the lowered (shipping) position. A gaseous nitrogen purge is applied to the oxidizer pump seal during the time the engine is being rotated to the horizontal or lowered position. The engine is then secured on Air Transport Engine Handler G4044 in the lowered position and the sling removed. If the engine is to be shipped cross-country by truck, the turbopump shaft preload fixture is installed. A check is then made to make sure that

Thrust Chamber Throat Security Closure G4089 is installed, that all desiccant is correctly secured, that the humidity range is acceptable, that openings are covered with suitable closures, and that the gimbal bearing is immobilized with Gimbal Bearing Lock G4059. The frame and Engine Cover G4047 are installed on the engine with the necessary forms sealed in the security pouch. Using a facility hoist and Engine Handler Sling G4052, the nozzle extension is installed on Nozzle Extension Handling Fixture G4080 and the loaded nozzle extension installed on Handling Adapter G4081. Because of shipping regulations governing transportation of ignition devices, the engine hypergol cartridge and pyrotechnic igniters are not shipped with the engine.

1-147. SHIPPING BY TRUCK.

1-148. Trucks are used to transport the engine, nozzle extension, and loose equipment, cross-country or to and from dock sites. Using a facility hoist and Engine Handler Sling G4052, the handler-installed engine and loaded nozzle extension (installed on the handling adapter) are loaded and secured on low-bed, air-ride-equipped trailer. Loose equipment is packaged in boxes, loaded by forklift, and secured. For cross-country shipping, a calibrated impact recorder is installed on the handler. A truck transport checklist is used as a guide to verify that specified procedures are performed before truck departure and during cross-country shipping.

1-149. SHIPPING BY SHIP.

1-150. The engine, nozzle extension, and loose equipment are delivered to the ship by truck. The low-bed trailer is positioned on the ship's deck. Using a mobile crane, Engine Handler Sling G4052, and tractor, the Handler-installed engine is removed from the trailer, placed on the cargo deck, then moved forward and secured. The nozzle extension and loose equipment are removed from the trailer by mobile crane or forklift and secured to the cargo deck. The water transport checklist is used as a guide to verify that specified procedures are performed before departure, in transit, and after docking.

1-151. RECEIVING ENGINE AT MAF.

1-152. The Stage Contractor receives the engine and is responsible for engine flow at MAF. Detailed requirements for engine receiving by truck and ship are in R-3896-9. Detailed requirements describing the use of engine handling equipment are in R-3896-3.

1-153. RECEIVING BY TRUCK.

1-154. Engines, nozzle extensions, and loose equipment received by cross-country truck or by truck from the MAF dock are delivered to the Manufacturing Building where the equipment is visually inspected for shipping damage. If arriving at MAF by cross-country truck, the arrival time and date are recorded

5-10 An F-1 turbopump in the final stages of being mounted on the side of this development engine's upper thrust chamber. The heat exchanger will connect the turbine at the bottom of the turbopump to the turbine exhaust manifold, as in the case of the engine in the background. (Rocketdyne, Vince Wheelock Collection)

on the impact recorder chart. Using the facility hoist and Engine Handler Sling G4052, the handler-installed engine and nozzle extension are moved from the trailer to the floor. Loose equipment is removed from the trailer using a forklift. The nozzle extension is routed to the nozzle extension storage area, and loose equipment is routed to the Engine Support Hardware Center. The engine is routed to the engine area or to the bonded storage area (if unassigned), where the impact recorder and turbopump preload fixture are removed (if installed) and returned to Canoga Park.

1-155. RECEIVING BY SHIP.

1-156. When the ship arrives at the MAF dock, a tug, mobile crane, and low-bed trailer are positioned on the ship's cargo deck for the off-loading procedure. Using Engine Handler Sling G4052 and the mobile crane, the engine and nozzle extension are loaded and secured on the low-bed trailer. The loose equipment is loaded on the trailer by forklift. The trailers are moved into the Manufacturing Building, where the engine, nozzle extension, and loose equipment are inspected for shipping damage. Engine receiving then proceeds as described in paragraph 1-153.

1-157. UNASSIGNED-ENGINE FLOW AT MAF.

1-158. Unassigned-engine flow at MAF pertains to unassigned and spare engines. Upon receipt at the Manufacturing Building, unassigned engines are inspected for shipping damage, moved to the bonded storage area, inspected, and stored until scheduled for modification and/or assigned to a stage. Spare engines are processed through buildup and single-engine check-out, moved to the bonded storage area, and stored in a standby condition in case engine replacement is required. Single-engine checkout is required for all engines in storage over six months. If any discrepancies are observed during engine flow at MAF, Engine Contractor personnel perform unscheduled maintenance and repair or replace discrepant hardware on the engine. Discrepant hardware removed from the engine is routed to the CM&R area,[2] where it is repaired and tested.

1-159. STORAGE RECEIVING INSPECTION.

1-160. Unassigned engines are visually inspected in the bonded storage area. The engine cover is removed, and the engine inspected for damage, corrosion, residual fluid on exterior surfaces, and surface wetting on the hydraulic control system exterior. It is verified that specified areas of the engine are coated with corrosion preventive, that humidity indicators indicate blue, and that line markings are correct. The turbopump preservation status is checked in the Engine Log Book, and the turbopump is serviced if required. The engine cover is reinstalled. Detailed inspection requirements for engines in storage are in R-3896-11.

1-161. ENGINE FLOW AT MAF.

1-162. When an uninstalled engine is received in the engine area, it is removed from Air Transport Engine Handler G4044, rotated to the vertical position, and placed on Engine Handling Dolly G4058 using Engine Rotating Sling G4050 and the facility hoist. The engine is then moved into a workstand where receiving inspection and engine buildup are accomplished. After engine buildup, the engine is placed into a test stand for single-engine checkout and installation of wrap-around lines. The engine is then removed from the test stand, rotated to the horizontal position and installed on Engine Handler G4069. The oxidizer pump

[2] The CM&R area is an environmentally controlled room set aside for component maintenance and repair of F-1 engines.

seal is purged with gaseous nitrogen during engine rotation to the horizontal position and for 30 minutes (minimum) thereafter. The engine is moved to the Stage Horizontal Final Assembly Area, where the engine is prepared for installation, installed in the stage, and inspected in preparation for shipment to MTF. Engine modifications are made as required during engine flow at MAF. If any discrepancies are observed, Engine Contractor personnel perform unscheduled maintenance, and repair or replace hardware on the engine. Discrepant hardware removed from the engine is routed to the CM&R area, where it is repaired and tested. Detailed requirements describing the use of engine handling equipment are in R-3896-3.

1-163. RECEIVING INSPECTION.

1-164. After installation in the single-engine workstand in the engine area of the Manufacturing Building, each assigned engine undergoes an overall visual receiving inspection. The engine is visually inspected for damage, corrosion, and missing equipment; for evidence of fluid in drain line exits or on the engine exterior; and for surface wetting on the hydraulic control system exterior. It is verified that corrosion preventive and aluminum-foil tape is present in specified areas, line markings are correct, humidity indicators indicate blue, and there are no voids in the turbopump housing cavity filler material. A clean polyethylene bag is installed on the fuel overboard drain line, the turbopump preload fixture is removed, and orifice sizes and serialized components are checked against those listed in the Engine Log Book. Detailed inspection requirements for engines received at MAF are in R-3896-11.

1-165. ENGINE BUILDUP, MODIFICATION, AND MAINTENANCE.

1-166. LOOSE EQUIPMENT INSTALLATION. Loose equipment that does not interfere with single-engine checkout is installed during engine buildup. The electrical cable support post is installed only on engines assigned to the outboard positions. The interface panel-to-oxidizer inlet insulation seal is installed on all engines. Wrap-around ducts and hoses are not installed at this time.

1-167. THERMAL INSULATION BRACKETRY INSTALLATION. The field-installed thermal insulation bracketry is normally stored at MAF until installation on the engine. All brackets are installed except for the bracket that attaches to the engine handling bearing. The engine handling bearing is an attach point for securing the engine onto Engine Handler G4069; therefore, the bracket is installed after the engine is installed on the stage. Requirements for installing thermal insulation brackets are in R-3896-6.

1-168. MODIFICATION AND MAINTENANCE. Modifications are made and maintenance tasks are performed during engine buildup, whenever possible. Engine modifications and special inspections consist of incorporating retrofit kits, as a result of Engineering Change Proposals (ECPs), and implementing Engine Field Inspection Requests (EFIRs). Engine maintenance involving component removal and replacement or turbopump disassembly, if required, is done

5-11 The gimbal bearing assembly was bolted to the top of the LOX dome assembly. (Rocketdyne, Harold C. Hall Collection)

in the engine workstands. Component modification, repair, and functional testing are done in the environmentally controlled CM&R area.

1-169. THRUST VECTOR CONTROL SYSTEM INSTALLATION. The thrust vector control system is installed by the Stage Contractor on engines assigned to the outboard positions. This system consist of two gimbal actuators, hydraulic supply and return lines, and a hydraulic filter manifold.

1-170. SINGLE-ENGINE CHECKOUT.

1-171. Single-engine checkout is done after receiving inspection and after engine buildup tasks are completed. The engine is installed in the test stand, where the ignition monitor valve sense tube is disconnected, Thrust Chamber Throat Security Closure G4089 removed, and Thrust Chamber Throat Plug G3136 installed. All connections are made between the engine and Engine Checkout Console G3142; facility electrical, pneumatic, and hydraulic sources are applied to the console; and the console is prepared for operation. Electrical system function and timing tests, a turbopump torque test, pressure tests, valve timing tests, and leak and function tests are done in accordance with the detailed requirements in R-3896-11. Upon completion of engine checkout, the ignition monitor valve sense tube is connected, Thrust Chamber Throat Plug G3136 is removed, and Thrust Chamber Throat Security Closure G4089 installed.

1-172. WRAP-AROUND DUCT AND HOSE INSTALLATION.

1-173. The loose-equipment wrap-around ducts and hoses are installed on the engine in the test stand after single-engine checkout. The helium, GOX, and hydraulic wrap-around ducts and the purge and prefill hoses are installed and connected to flanges used for test setups during engine testing. The ducts and hoses are aligned using alignment tool T-5041233 and supported with support set T-5046440 to prevent movement until the engine is installed in the stage and interface connections are completed. The engine is then removed from the test stand. Detailed requirements for installing and aligning wrap-around ducts and hoses are in R-3896-3.

1-174. ENGINE INSTALLATION AT MAF.

1-175. PREPARATION FOR ENGINE INSTALLATION. The engine is rotated to the horizontal position and installed on Engine Handler G4069 using Engine Rotating Sling G4050 and the facility hoist. The oxidizer pump seal is purged during engine rotation to the horizontal position and for 30 minutes (minimum) thereafter. After removing the interface panel access doors, the oxidizer and fuel inlet covers are removed, the inlets inspected for contamination, the oxidizer inlet screen and seal secured in place, and the inlets covered with Aclar film. The fuel overboard drain system is isolated using clean polyethylene bags. The gimbal boot cover is removed, and it is verified that the gimbal bearing locks are installed, the electrical cable support posts installed on engines assigned to outboard positions, and the engine gimbal wrap-around lines are installed and adequately supported. When ready for installation in the stage, the engine is moved to the Stage Horizontal Final Assembly Area and positioned under a mobile hoist. (A-frame). Thrust Chamber Throat Security Closure G4089 is removed and the thrust chamber inspected. The engine horizontal installation tool is suspended from the mobile hoist, prepared for engine installation, and then installed in the thrust chamber. The engine is then removed from Engine Handler G4069 and raised and rotated in the position required for engine installation.

Detailed requirements for fuel overboard drain system isolation and engine preparation for installation are in R-3896-11.

1-176. ENGINE INSTALLATION. When preparations for engine installation are completed and the engine is correctly positioned in the stage, the engine gimbal bearing is mated and secured to the stage attach point. On the outboard engines, the gimbal actuators are secured to the stage actuator locks, while on the inboard engine, the stiff arms are secured to the actuator locks. Gimbal bearing locks are removed, and the gimbal boot is reinstalled on the gimbal bearing. The engine horizontal installation tool is removed from the thrust chamber after the engine is secured to the stage; then the Thrust Chamber Throat Security Closure G4089 is installed. Aclar film is removed from engine oxidizer and fuel inlets, fuel inlet seals and screens are installed, and stage ducting is connected to the engine inlets. The interface electrical connectors and stage pressure switch checkout supply line are connected at the interface panel, and the wrap-around ducts and hoses are connected to the stage. The thermal insulation bracket that attaches to the engine handling bearing is installed as specified in R-3896-6. Detailed requirements for installing the engine are in R-3896-11.

1-177. MANUFACTURING INSTALLATION VERIFICATION. When engine installations and stage assembly are completed, the Stage Contractor performs a manufacturing installation verification. This verification consists of a gaseous nitrogen leak test of the engine interface connections and stage systems.

1-178. INSTALLED-ENGINE INSPECTION BEFORE STAGE SHIPMENT TO MTF.

1-179. The installed-engine inspection before shipment to MTF is made after the stage assembly and verification tests are complete. Each engine is visually inspected for damage, corrosion, and missing equipment; for evidence of fluid in drain line exits, fluid on the engine exterior; and for surface wetting on the hydraulic control system exterior. It is verified that corrosion preventive and aluminum-foil tape is present in specified areas, line markings are correct, the humidity indicator in the thrust chamber throat security closure indicates blue, and there are no voids in the turbopump housing cavity filler material. The fuel overboard drain system isolation polyethylene bags are visually inspected for fluid. If fluid is present, the bags are emptied and the quantity of fluid is measured. The turbopump preservation status is checked in the Engine Log Book, and the turbopump is serviced if required. A final updating of the Engine Log Book is made before engine shipment to MTF. Detailed procedures for inspecting the installed engine before shipment to MTF are in R-3896-11.

1-180. STAGE SHIPMENT TO MTF.

1-181. When installed-engine inspection is complete, the forward stage cover and engine covers are installed, the workstands and platforms are rolled away from the engines, a tractor is connected to the stage transporter, and the stage is pulled to the MAF dock. The stage is loaded onto the barge and secured. The nozzle

This book is dedicated to all my “favorites”.

You know who you are.



KMH Publishing



KMHPublishing@yahoo.com

ISBN -13: 978-0615973289

ISBN -10: 0615973280

# Frank, The Bull Ridin' Frog

Written and Illustrated by

Karen M. Hines

Frank’s hat was black

and his belly was yellow.

He wore big boots

for such a little fellow.

Frank lived on a ranch

out in the Wild West.

A ‘Champion’ cowboy

who always tried his best.

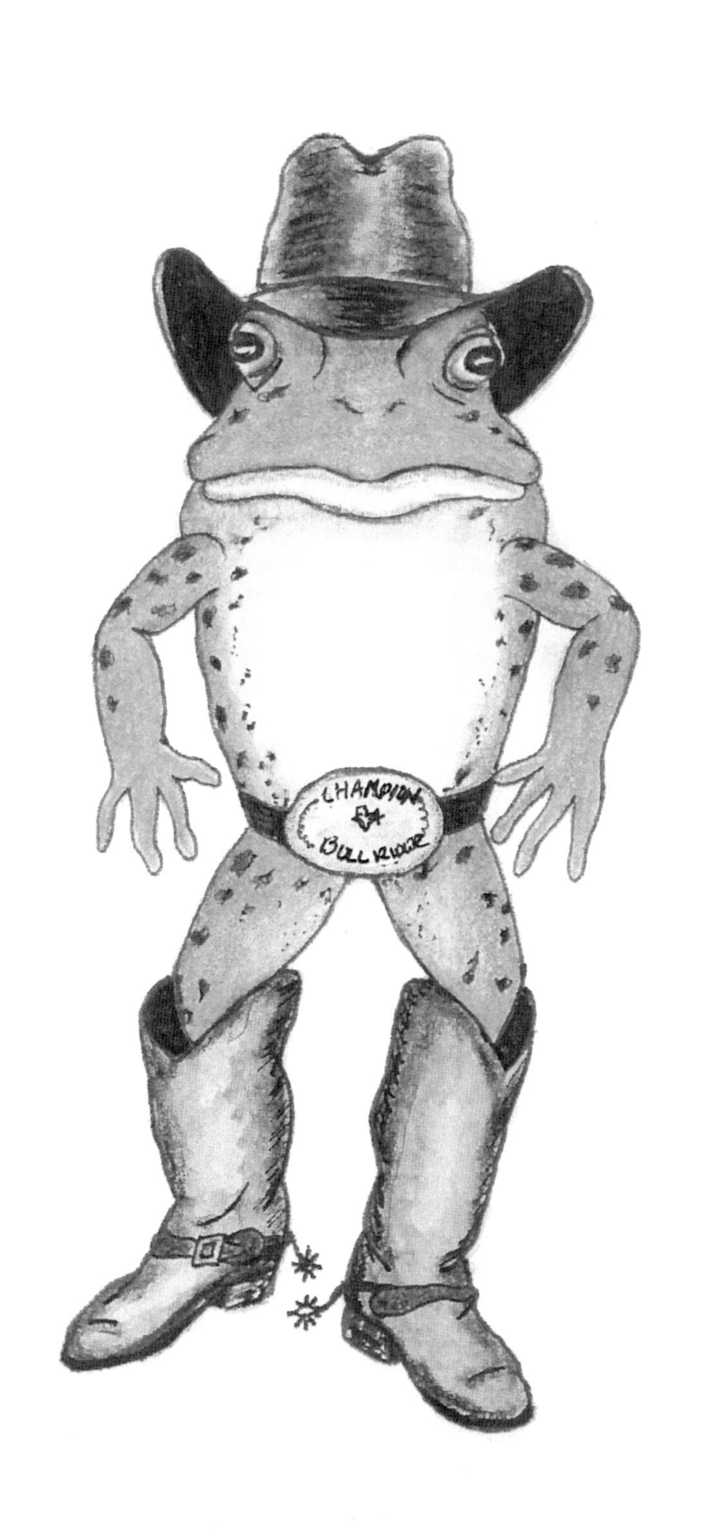
CHAMPION

"Wibbit, wibbit, wibbit,"

young Frank would hop and say.

The other frogs would laugh at him

when they went out to play.

"He can't even say r..r..r..ribbit!"

the little frogs would joke.

Fun at Frank

they liked to poke.

It made Frank sad

that he wasn't like the rest.

Having a friend to like him

would be the very best.

Grandma smiled and said,

"Keep on trying,

you're doing great so far.

But remember I love you

just the way you are."

One day the Frog Rodeo rolled into town.

Frogs  gathered to watch from ranches all around.

Frank hopped on the fence and watched the cowboys ride and fall.

He dreamed of roping and riding.

Frank wanted to do it all.

DYLAN
FRANK

Frank tried to rope a mosquito

as it went buzzing by

but his little tongue wasn't

fast enough.

"I'll never do it," he did cry.

Grandma smiled and said,

"Keep on trying,

you're doing great so far.

But remember I love you

just the way you are."

"Wibbit, wibbit, wibbit,"

Frank would hop and say

as he worked on his cowboy skills

getting  better every day.

Frank practiced every day.

He tried, and tried, and tried.

"He's a pretty good cowboy,"

the old frogs sighed.

Grandma smiled and said, “Keep on trying, you’re doing great so far. But remember I love you just the way you are.”

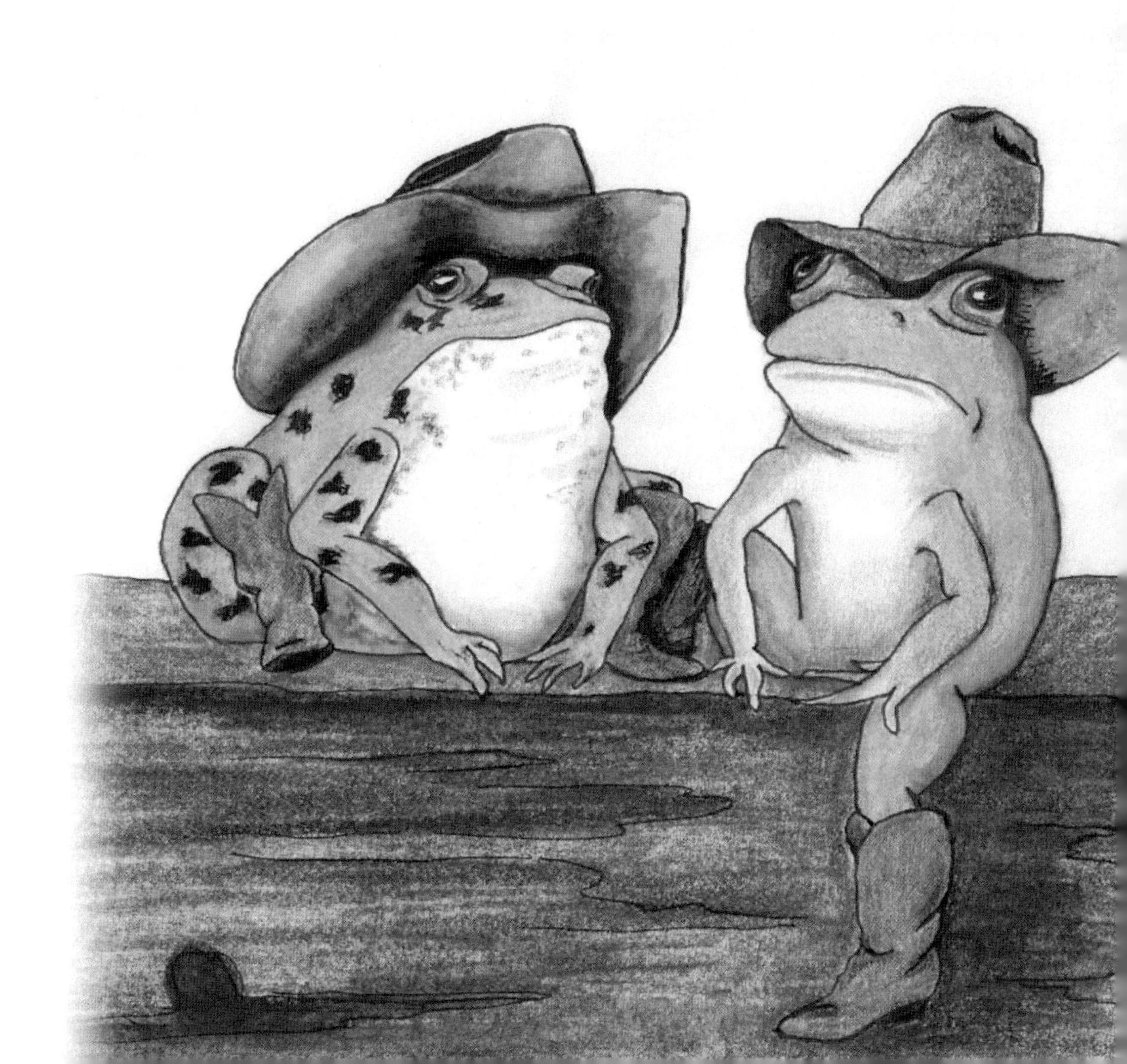

Soon, Frank roped a fly

flying by in the sky.

Tied up its legs 1...2...3,

faster than you can blink an eye!

Frank could ride
a bull better than
any frog in the West
Frank practiced very
hard and always
tried his best.

The Frog Rodeo was back in town
and Frank was ready to ride.

Horses, bulls, boots and spurs,

"Yeee-haaaaww!!" the cowboys cried.

The bulls were bigger

than any Frank had ever seen.

They hissed and spit and

pawed the ground.

Boy-oh-boy did they look mean!

Cowboys went flying

as the bulls left the chutes.

Frank's turn was next....

he swallowed hard....

and started shaking in his boots.

Grandma smiled and said,

"Keep on trying,

you're doing great so far.

But remember I love you

just the way you are."

Frank hopped up on that big,

mean bull and it began to twist.

When it started to buck and kick

he held on tight with his tiny fist.

BUZZZZZZZZZZZZZZZ!!!

went the eight second timer

and the crowd began to cheer!

Frank rode the bull!

Frank was the best!

The “Champion Bull Rider” that year.

But then……

That bull gave one final buck.

Frank flew up in the air!

Over and over;

his boots flying here and there!

“W-W-W-I-I-B-B-B-I-I-T-T-T!!!”

Frank was heard to yell.

Then with a loud thud...

Frank fell.

"W-W-W-I-I-B-B-B-I-I-T-T-T!!!"

Frank didn't move.

He just laid in the dirt.

The other frogs screamed,

"Is he hurt? Is he hurt?"

Oh No!!

Then Frank sat up

at a slow...slow...pace

and a great big smile

was on his face!

Frank never quit trying.

He was a champion by far.

The crowd yelled,

“FRANK!

WE LOVE YOU!

JUST THE WAY YOU ARE!”

...and Grandma smiled.

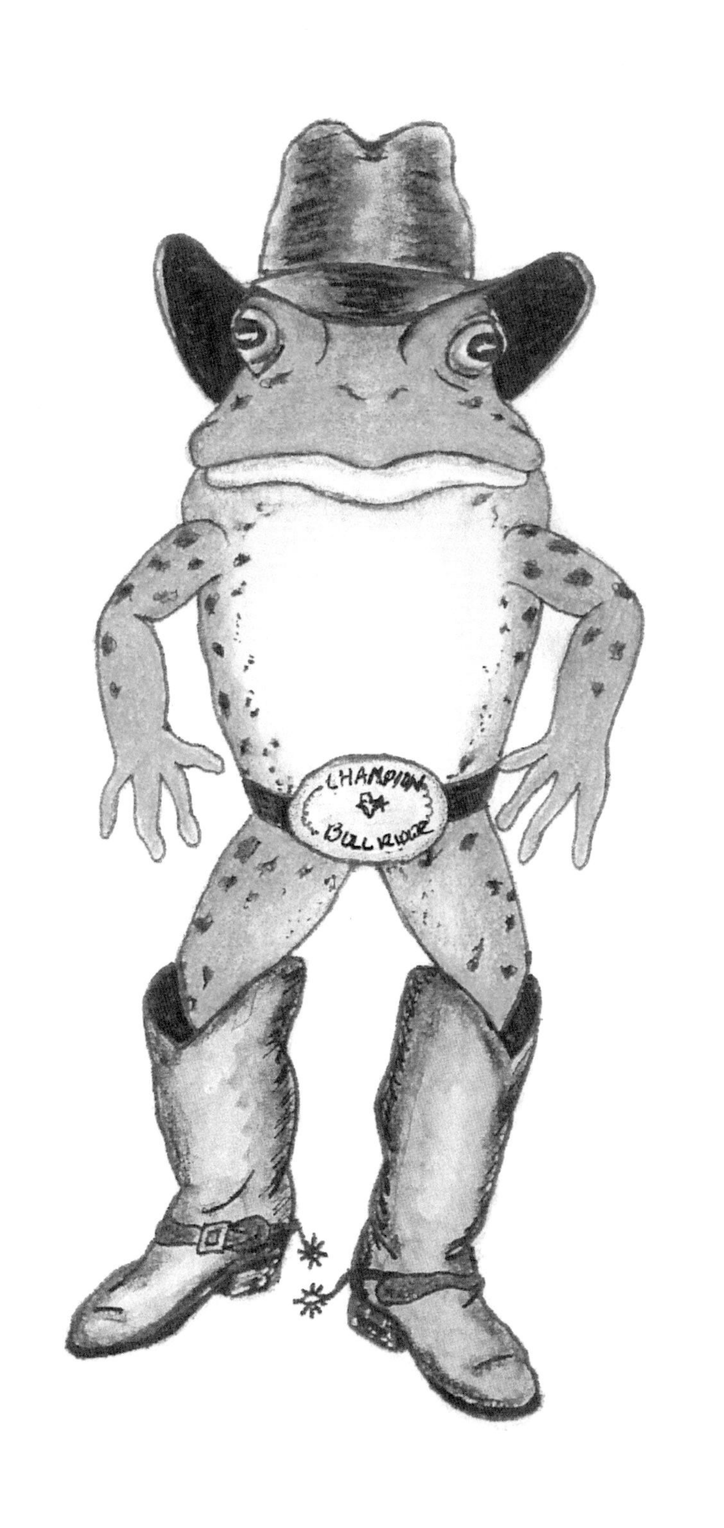
CHAMPION
BULL RIDER

That’s The End Y’all!

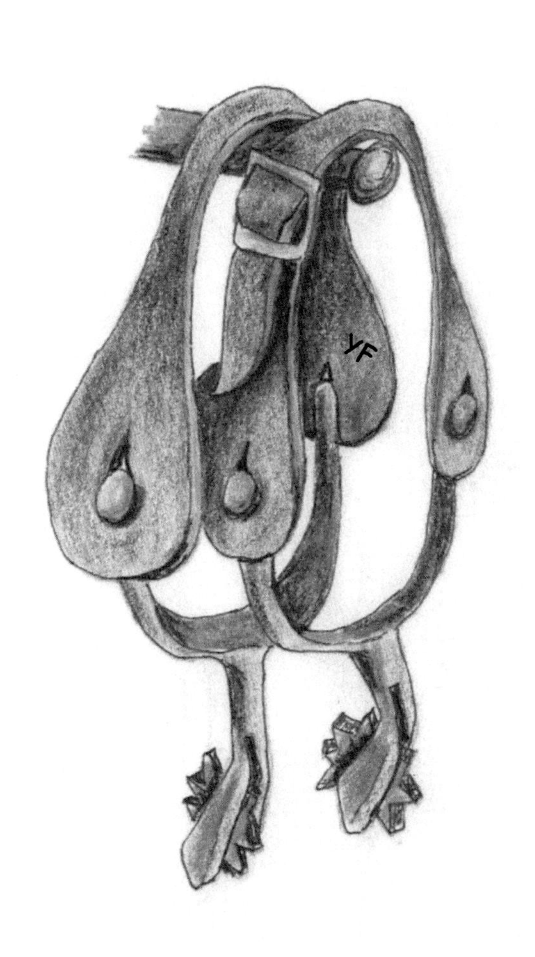
YF

Made in the USA
Las Vegas, NV
04 September 2021

5-12 Rocketdyne built new rocket engine assembly buildings in the mid-1960s that included areas for vertically assembling F-1 engines. (Rocketdyne, Harold C. Hall Collection)

extensions are loaded on low-bed trailers, towed to the MAF dock, loaded on the barge using a mobile hoist, and secured. The barge is then moved to MTF by tug.

1-182. STAGE FLOW AT MTF.

1-183. The stage is received at MTF and installed in the test stand. The engine covers are removed, and receiving inspection is performed. The nozzle extensions, slave hardware (normally stored at MTF), and MTF static test instrumentation are installed; then a pre-static checkout is performed. Thermal insulation is not required for static test, therefore it is not installed. Engine maintenance is done and modifications are made as required during engine flow at MTF. Upon completion of pre-firing preparations, the static firing test is performed. After static test, the engines are inspected; the test instrumentation, slave hardware, and nozzle extensions are removed; a pre-shipment inspection is performed; and the stage and nozzle extensions are removed from the test stand and loaded on the barge for return to MAF.

1-184. STAGE INSTALLATION IN TEST STAND.

1-185. When the stage arrives at MTF, the barge is docked next to the test stand. Test stand overhead cranes are attached to the forward and aft ends of the stage;

the stage is lifted clear of the stage transporter and barge, rotated to the vertical position, and positioned into the test stand. During rotation to the vertical position, the thrust chamber and exhaust manifold are monitored for fuel leakage. The stage is secured to the test stand with mechanical holddowns; stage/facility propellant, hydraulic, pneumatic, and electrical connections are secured; and engine covers and engine oxidizer and fuel inlet screens are removed.

1-186. ENGINE RECEIVING INSPECTION.

1-187. After the stage is installed in the test stand, the engines undergo an overall visual receiving inspection. Each engine is inspected for damage, corrosion, and missing equipment and for evidence of fluid in drain line exits. It is verified that corrosion preventive and aluminum-foil tape is present in specified areas, the engine soft goods installed life is within specified limits, and there are no voids in the turbopump housing cavity filler material. The fuel overboard drain system isolation polyethylene bags are visually inspected for fluid. If fluid is present, the bags are emptied and the quantity of fluid is measured. Engine orifice sizes and serialized components are checked against those listed in the Engine Log Book. Detailed inspection requirements for installed engines received at MTF are in R-3896-11.

1-188. INSTALLATION OF NOZZLE EXTENSIONS, SLAVE HARDWARE, AND MTF STATIC TEST INSTRUMENTATION.

1-189. The nozzle extensions, slave hardware, and MTF static test instrumentation are installed on the engines after the stage is installed in the test stand and after receiving inspection. Using Engine Handler Sling G4052 and overhead cranes, the nozzle extension is removed from the barge and from Nozzle Extension Handling Fixture G4080 and Handling Adapter G4081 and placed on Engine Vertical Installer G4049 on the lower stand work platform. The installer, with nozzle extension, is positioned below the engine; then the nozzle extension is installed on the engine, and the installer lowered. The polyethylene bags are removed from the fuel overboard drain system, and the slave fuel, oxidizer, and nitrogen overboard drain lines are installed. The slave igniter harness and MTF static test instrumentation are then installed and connected. Detailed installation requirements are in R-3896-11. Detailed nozzle extension handling requirements are in R-3896-9.

1-190. STAGE PRE-STATIC CHECKOUT.

1-191. The stage pre-static checkout is performed on all engine and stage systems. Immediately preceding pre-static checkout, Thrust Chamber Throat Security Closure G4089 is removed and Thrust Chamber Throat Plug G3136 is installed. The checkout consists of electrical, hydraulic, and pneumatic leak and function tests. A simulated static test, which simulates stage preparation, engine start, ignition, mainstage, and cutoff sequencing, is performed to verify stage acceptability for static test. Detailed pre-static checkout requirements are in R-3896-11.

1-192. STATIC TEST

1-193. When all required checkout procedures, modifications, and maintenance are completed, and the Thrust Chamber Throat Plug G3136 is removed, the hypergol cartridge and pyrotechnic igniters are installed and checked out and the test area is cleared in readiness for static test. A 125-second, uninterrupted-duration stage static test is made to checkout all electrical-electronic, propulsion, mechanical, pressurization, propellant, and control systems that function during actual countdown, launch, and flight. Measurements of the static test are recorded and processed to determine stage acceptability and flight readiness. The engine start for the stage is a 1-2-2 sequence: the center engine starts first, and the remaining outboard engines start in opposed groups of two. The engine cutoff is a 3-2 sequence: the center engine and two opposite outboard engines cut off first; then the remaining two outboard engines cut off.

1-194. ENGINE INSPECTION AFTER STATIC TEST.

1-195. The engine and nozzle extension are inspected visually after static test to verify that damage did not occur during the test. Detailed inspection requirements are in R-3896-11 and include inspecting for exterior damage and missing aluminum tape between thrust chamber exhaust manifold and thrust chamber tubes; inside of thrust chamber for tube and injector damage, injector contamination, and liquid leakage. Other inspections are for tension tie deformation, bent or broken studs, nozzle extension for carbon deposits around flange area, and internal damage and erosion.

1-196. STATIC TEST DATA REVIEW.

1-197. The static test data is reviewed after static test to determine that the engine is operating within specified limits. Test instrumentation readings are examined to detect abnormalities, sudden shifts, oscillations, or performance near the minimum or maximum limits.

1-198. TURBOPUMP PRESERVATION.

1-199. The turbopump is preserved within 72 hours after static test. After removing fluid through the turbopump No. 3 bearing drain line, the turbopump bearings are purged with gaseous nitrogen, and five gallons of preservative oil is supplied to the bearings while the turbopump is slowly rotated. The fluid is then drained through the No. 3 bearing drain line, and the bearings are again purged with gaseous nitrogen. The preservation date is recorded in the Engine Log Book.

1-200. REMOVAL OF NOZZLE EXTENSIONS, SLAVE HARDWARE, AND MTF STATIC TEST INSTRUMENTATION.

1-201. Engine Vertical Installer G4049 is positioned below the nozzle extension and the nozzle extension removed from the engine and lowered onto the installer. Using Engine Handler sling G4052 and overhead cranes, the nozzle extension is removed from the installer, installed on Nozzle Extension Handling Fixture

5-13 The thermal insulation for each F-1 engine was verified in the final stages of assembly, then removed for engine shipment to NASA. This photograph was taken November 4, 1966. (Rocketdyne, Harold C. Hall Collection)

G4080, and the loaded nozzle extension installed on Handling Adapter G4081. The slave hardware, consisting of fuel overboard drain lines and the igniter harness, is removed, cleaned, tested, and repaired or replaced, as required, for reuse during the next static test. The fuel overboard drain system is isolated using clean polyethylene bags. The expended igniters and hypergol cartridge are removed. The MTF static test instrumentation is disconnected and removed and the instrumentation ports plugged immediately by incorporating the applicable retrofit kit specified in Modification Instruction R-5266-391 (ECP F1-391). The Thrust Chamber Throat Security Closure G4089 is installed. Detailed removal requirements are in R-3896-11. Detailed nozzle extension handling requirements are in R-3896-9.

1-202. INSTALLED-ENGINE INSPECTION BEFORE STAGE SHIPMENT TO MAF.

1-203. The engine is inspected before shipment to MAF and after all post-static-test tasks are complete. Each engine is visually inspected for damage, corrosion, and missing equipment; for evidence of fluid in drain line exits or on the engine exterior; and for surface wetting on the hydraulic control system exterior. It is verified that corrosion preventive and aluminum-foil tape is present in specified areas, line markings are correct, the humidity indicator in the thrust chamber throat security closure indicates blue, and there are no voids in the turbopump housing cavity filler material. The fuel overboard drain system isolation polyethylene bags are visually inspected for fluid. If fluid is present, the bags are emptied and the quantity of fluid is measured. All engine protective closures are installed upon completion of visual inspection. It is verified that the humidity indicator in the thrust chamber throat security closure indicates blue at the time of shipment. Detail inspection requirements are in R-3896-11.

1-204. STAGE REMOVAL FROM TEST STAND.

1-205. After engine visual inspection, the engines and stage are prepared for removal from the test stand. The engine and stage covers are installed; stage/facility propellant, hydraulic, pneumatic, and electrical connections are disconnected; and mechanical holddowns are removed. Test stand overhead cranes are attached to the forward and aft ends of the stage; the stage is lifted clear of the test stand, rotated to the horizontal position, and installed on the stage handler on the barge. The oxidizer pump seal is purged during engine rotation to the horizontal position for 30 minutes (minimum) thereafter. The nozzle extensions, installed on Nozzle Extension handling Fixtures G4080 and Handling Adapters G4081, are removed by overhead crane and loaded on the barge. The stage transporter and nozzle extensions are secured on the barge for shipment. A final updating of the Engine Log Book is made before shipment to MAF.

1-206. STAGE SHIPMENT TO MAF.

1-207. The barge, containing the stage and nozzle extensions, is moved from MTF to MAF by tug. Upon arrival at the MAF dock, a tractor is connected to the stage transporter, and the stage is pulled from the barge and towed to the Stage Checkout Building. The nozzle extensions are loaded on low-bed trailers, using a mobile hoist, and towed from the barge to the nozzle extension storage area.

1-208. STAGE FLOW AT MAF.

1-209. The stage is positioned in the Stage Checkout Building at MAF, and workstands and platforms are installed to aid access during inspection and checkout. The engines undergo a receiving inspection, refurbishment, post-static checkout, and pre-shipment inspection. A storage period may be required after refurbishment, if so, the stage is prepared for storage and stored for a specified time before post-static checkout.

1-210. ENGINE RECEIVING INSPECTION.

1-211. After positioning the stage in the Stage Checkout Building, the engines undergo an overall visual receiving inspection. Each engine is inspected for damage, corrosion, and missing equipment and for evidence of fluid in drain line exits. It is verified that corrosion preventive and aluminum-foil tape is present in specified areas and that there are no voids in the turbopump housing cavity filler material. The fuel overboard drain system isolation polyethylene bags are visually inspected for fluid. If fluid is present, the bags are emptied and the quantity of fluid is measured. Engine orifice sizes and serialized components are checked against those listed in the Engine Log Book. It is verified that the humidity in the thrust chamber throat security closure indicates blue. Detailed inspection requirements for installed engines received MAF are in R-3896-11.

1-212. ENGINE REFURBISHMENT.

1-213. The engine is refurbished after receiving inspection. The engines are first cleaned of any foreign matter and corrosion that may have resulted from exposure to rain, humidity, sand, or dust. The oxidizer dome insulator is installed in accordance with requirements specified in R-3896-6. The flight igniter harness is installed, tested and connected in accordance with requirements specified in R-3896-11. Outstanding maintenance or modification, as required by ECPs and EFIRs, is done during the refurbishment period.

1-214. STAGE STORAGE.

1-215. Storage of installed engines is scheduled following completion of refurbishment. The amount of time the stage remains in storage is determined by the Saturn V vehicle launch schedule. Stage storage, in excess of six months, requires that engine post-static checkout be performed when the stage is removed from storage. Installed engines are visually inspected for damage, corrosion, and missing equipment, and for evidence of fluid in oxidizer and nitrogen purge overboard drain lines. It is also verified that corrosion preventive and aluminum-foil tape is present in specified areas, the gimbal boot is installed, there are no voids in the turbopump housing cavity filler material, and that fuel overboard drain system isolation polyethylene bags do not contain fluid. If fluid is present, the bags are emptied and the quantity of fluid is measured. The turbopump preservation status is checked in the Engine Log Book and the turbopump is serviced if required; desiccants are installed in the thrust chamber throat security closure and the closure is installed; and humidity indicators are checked for a blue indication. The engine-to-stage gimbal actuators are locked to prevent engine movement, and the stage is stored in an environmentally controlled area. The engines are inspected periodically during storage. Detailed inspection requirements for installed engines in storage are in R-3896-11.

1-216. POST-STATIC CHECKOUT.

1-217. The post-static checkout is done after refurbishment tasks are completed, after a stage is removed from storage on which a post-static checkout had not been previously accomplished, or after stage storage has exceeded six months. The post-static checkout consists of complete electrical, hydraulic, and pneumatic leak and functional tests of the installed engines and stage systems. The post-static checkout is completed with a simulated launch test that consists of stage preparations, engine start, ignition, mainstage, liftoff, flight, and engine cutoff in the prescribed sequence to assure flight readiness of the engines and stage. Post-static

5-14 F-1 engine No. 2088 prior to mechanical and electrical checkout in December 1968. (Rocketdyne, Harold C. Hall Collection)

checkout includes a flight instrumentation function test, turbopump torque test and heater function test, lead and function test of the bearing coolant control valve, hypergol manifold, thrust-OK pressure switches, thrust chamber prefill line, ignition monitor valve, oxidizer dome and gas generator oxidizer injector purge system, cocoon purge system, and hydraulic system. Leak test of the thrust chamber, heat exchanger helium and oxidizer systems, propellant fuel and oxidizer systems, exhaust system, and valve timing function tests are also accomplished. Installed engine tests are conducted in accordance with requirements specified in R-3896-11.

1-218. INSTALLED-ENGINE INSPECTION BEFORE STAGE SHIPMENT TO KSC.

1-219. The installed engine is inspected before shipment to KSC and the Engine Log book is reviewed after post-static checkout tasks are completed. Each engine is visually inspected for damage, corrosion, and missing equipment; for evidence of fluid in drain line exits, fluid on the engine exterior; and for surface wetting on the hydraulic control system exterior. It is verified that corrosion preventive and aluminum-foil tape is present in specified areas, that line markings are correct, that the humidity indicator in the thrust chamber throat security closure indicates blue, and that turbopump housing cavity filler material does not contain voids. The fuel overboard drain system isolation polyethylene bags are visually inspected for fluid. If fluid is present, the bags are emptied and the quantity of fluid is measured. The turbopump preservation status is checked in the Engine Log Book, and the turbopump is serviced if required. A final updating of the Engine Log book is made before engine shipment to KSC. Detailed procedures for inspecting the installed engine before shipment to KSC are in R-3986-11.

1-220. STAGE SHIPMENT TO KSC.

1-221. After the engine pre-shipment visual inspection is completed, the forward and aft stage covers are installed, workstands and platforms removed, and the stage pulled from the Stage Checkout Building to the MAF dock for transport to KSC by barge. The nozzle extensions, engine loose equipment, and the thermal insulation are loaded on low-bed trailers and transported to the MAF dock where they are removed from the trailers and loaded on the barge and secured for shipment. Handling requirements for nozzle extensions and loose equipment are in R-3896-9. After the nozzle extensions, loose equipment, and thermal insulation boxes are loaded and secured, the stage is loaded onto the barge and secured. The barge is then moved to KSC by tug.

1-222. STAGE FLOW AT KSC.

1-223. The barge arrives at KSC dock where the stage, nozzle extensions, loose equipment, and thermal insulation boxes are off-loaded. The stage is towed from the dock to the Vertical Assembly Building (VAB). The nozzle extensions, loose equipment, and thermal insulation boxes are loaded on low-bed trailers and transported to the VAB. The stage is removed from the stage transporter and erected onto the Launch Umbilical Tower (LUT) where the engine visual receiving inspection, loose equipment installation, modification and maintenance, stage and engine leak and functional tests, and thermal insulation installations are accomplished. These tasks are conducted concurrently with the Saturn V vehicle assembly and testing. A final updating of the Engine Log Book is made after engine activities during stage flow are complete.

1-224. STAGE INSTALLATION ONTO LAUNCH UMBILICAL TOWER (LUT).

1-225. The stage is received in the low bay of the VAB. The forward and aft stage covers are removed and the stage and engines prepared for rotation and installation onto the LUT. The Engine Service Platform (ESP) and the LUT are moved into the high bay. The stage, on the transporter, is moved from the transfer aisle to the erection bay where the stage is removed from the transporter and rotated to the vertical position by overhead cranes. The stage is then moved by high bay crane and erected on the LUT and secured with four mechanical holddowns. The ESP and LUT level platforms are positioned around the engines for receiving inspection.

1-226. ENGINE RECEIVING INSPECTION.

1-227. After the stage is installed onto the LUT, protective closures are removed and the engines undergo an overall visual receiving inspection. The engines are inspected to verify that damage did not occur during shipping and that all equipment listed on shipping documentation was received. Each engine is inspected for damage, corrosion, and missing equipment; for evidence of fluid in drain line exits, fluid on the engine exterior, and for surface wetting on the hydraulic control system exterior. It is verified that corrosion preventive and aluminum foil tape is present in specified areas, the engine soft goods installed life is within specified limits, there are no voids in the turbopump housing cavity filler material, and that turbopump and outrigger arm surfaces do not contain scratches through paint. The fuel overboard drain system insulation polyethylene bags are visually inspected for fluid. If fluid is present, the bags are emptied and the quantity of fluid is measured. Engine orifice sizes and serialized components are checked against those listed in the Engine Log Book. Oxidizer and fuel high-pressure duct covers and thrust chamber covers are installed after visual inspection completion. Detailed inspection requirements for installed engines received at KSC are in R-3896-11.

1-228. LOOSE EQUIPMENT INSTALLATION.

1-229. The engine loose equipment is installed after engine receiving inspection is completed. The loose equipment consists of the nozzle extension, oxidizer overboard drain line, fuel overboard drain line, nitrogen purge overboard drain line, and fuel inlet elbow-to-interface boots. Using Engine Handler Sling G4052 and overhead cranes, the nozzle extension is removed from Nozzle Extension Handling Fixture G4080 and Handling Adapter G4081 and placed on the Nozzle

5-15 Protective covers were placed over the threaded attachments used to secure the thermal insulation. (Rocketdyne, NASA)

Extension Installer. The five nozzle extensions and Nozzle Extension Installers are placed on the Engine Service Platform in their respective engine positions. The Engine Service Platform is then raised from ground level up through the opening in the LUT until the nozzle extension flanges are approximately 5 inches below the thrust chamber exit flanges. Final adjustments are made and the mating of the extension flanges to the thrust chamber exit flanges is done with the individual Nozzle Extension Installers. After the nozzle extensions are secured to the engines, the overboard drain lines are attached and secured. Loose equipment is installed in accordance with requirements specified in R-3896-11. Detailed nozzle extension handling requirements are in R-3896-9. The stage fins and engine shrouds are installed in accordance with stage contractor requirements.

1-230. MODIFICATION AND MAINTENANCE.

1-231. The engine modifications and special inspections may be made and maintenance tasks may be performed, if required, throughout the stage flow at KSC. Modifications and special inspections are made as a result of approved ECP or EFIR action, and scheduled through joint agreement between the customer, stage contractor, and engine contractor. The engine maintenance is performed, if required, as a result of discrepant hardware noted during receiving inspection or engine lead and functional testing.

1-232. STAGE FUNCTIONAL TEST.

1-233. The stage functional testing is started after stage installation onto the LUT. The electrical, hydraulic, and pneumatic lead and functional tests are made in conjunction with vehicle assembly. The stage functional test consists of a flight instrumentation function test, turbopump torque test and heater function test, engine sequence verification test, leak and function test of the bearing coolant system, hypergol manifold, thrust-OK pressure switches, thrust chamber prefill line, ignition monitor valve, oxidizer dome and gas generator oxidizer injector purge system, oxidizer pump seal purge system, cocoon purge system, and hydraulic system. A leak test of the thrust chamber, heat exchanger helium and oxidizer systems, propellant fuel and oxidizer systems, exhaust system, and valve timing function tests is also performed. Installed engine tests are performed in accordance with requirements specified in R-3896-11.

1-234. THERMAL INSULATION INSTALLATION.

1-235. The thermal insulation (TIS) is installed after engine leak and functional testing is complete. The TIS is installed to completely envelop the engine and provide protection from extreme temperatures created by plume radiation and backflow during cluster engine flight. To allow access for verifying the integrity of engine components and systems and to prevent possible insulator damage from fluid spillage, the TIS is not installed until engine testing is complete. The required sequence and methods for TIS installation is in R-3896-6. After the thermal insulation is installed and before moving the Saturn V vehicle from the VAB, an engine environmental cover is installed on each S-IC engine, from the thrust chamber throat area to the exit end of the nozzle extension, to protect the thermal insulation from inclement weather. The cover is wrapped around the thrust chamber and nozzle extension and placed so that engine overboard drain lines are exposed through holes provided in the cover, and access flaps, four places, are located to provide access to drain ports and igniters. Overlapping edges of the cover are laced together, excess material is gathered around the thrust chamber throat and folds tied, and the cover drawn tight under exit end of nozzle extension. Detailed requirements for installation of the cover are in R-3896-11.

1-236. SATURN V VEHICLE FLOW AT KSC.

1-237. While the S-IC Stage is being received and erected in the VAB, the S-II Stage, S-IVB Stage and Instrument Unit are received in the VAB and placed in the checkout bays where they undergo a complete pre-erection checkout. Upon completion of S-IC erection, the Saturn V vehicle assembly is started, concurrently with S-IC Stage testing. When the fins, fairings engine shrouds and nozzle extensions are installed, the S-IC Stage assembly is complete. The Instrumentation Unit is moved into the high bay, placed on a platform near the S-IC Stage, and an S-IC Stage-Instrumentation Unit-checkout is performed.

Upon completion of pre-erection checkout, the S-II Stage is moved from the checkout bay to the high bay and mated with the S-IC Stage. The S-IVB Stage is moved from the checkout bay and mated with the S-II Stage, and the Instrument Unit is removed from the platform and mated with the S-IVB Stage, completing the assembly of the Launch Vehicle (LV). After individual modules are checked out at the Manned Spacecraft Operations Building (MSOB), the Apollo spacecraft, consisting of the mated lunar excursion, and service and command modules, is moved into the VAB and mated mechanically (lunar excursion module-adapter to forward mating flange of the instrument unit).

1-238. VEHICLE TESTING.

1-239. After the Apollo spacecraft and launch vehicle are mechanically mated, spacecraft modules are connected to their umbilicals from the umbilical tower of the mobile launcher and pre-power-on tests are made. When it has been determined that all flight and ground systems are satisfactory, full power is applied to the spacecraft. The spacecraft is then mated electrically to the launch vehicle and combined system tests, consisting of simulated countdowns and flights that exercise both flight and ground systems, are made. During the final combined system testing phase, the spacecraft and launch vehicle ordnance, minus pyrotechnics, are installed including the launch escape system. When the combined system testing is complete, the test data is reviewed, and if acceptable, the Saturn V vehicle is ready to be moved to the launch pad.

1-240. TRANSFERRING VEHICLE TO LAUNCH PAD.

1-241. The Apollo/Saturn V is transported from the VAB to the launch pad by the crawler transporter. The extendable platforms that enclosed the vehicle in the VAB are retracted, connections between the mobile launcher terminals and the terminals in the high bay are disconnected, the doors of the high bay are opened, and the transporter brought in and positioned beneath the platform section of the launcher. Hydraulic jacks are extended from the transporter to lift the launcher clear of its pedestals. Then, at a speed of approximately 1 mph, the transporter carries the launcher and the fully assembled Apollo/Saturn V to the launch pad for positioning.

1-242. LAUNCH PREPARATIONS AND TESTING.

1-243. After all electrical and pneumatic lines to the Apollo/Saturn V are reconnected through terminals at the base of the mobile launcher, and propellant lines, also connected through the launcher, are verified as correct, and it has been ascertained that no changes have occurred in the vehicle since it left the VAB, tests are made on the communication links to the vehicle. Measurements are also taken on systems such as the cutoff abort unit, radio-frequency, tank pressurization, and launch vehicle stage propellant utilization system. A Flight Readiness Test (FRT), backup guidance system test, and S-IC fuel jacket/oxidizer dome flush and purge are performed. Hypergolic propellants are loaded in the spacecraft tanks, RP-1 fuel is loaded in the launch vehicle tanks, and Countdown

5-16 F-1 engine No. 007 during a leak and functional checkout on May 3, 1962. Note the configuration of the turbine exhaust manifold. (Rocketdyne, Frank Stewart Collection).

Demonstration Tests (CDDT) are performed. Liquid oxygen and liquid hydrogen are loaded into the launch vehicle during the last few hours of the countdown.

1-244. SATURN V VEHICLE LAUNCH.

1-245. The data in this paragraph is only used to describe a typical vehicle launch and is not intended to represent actual launch data. With S-IC stage engines and launch vehicle preparations complete, the S-IC engines are fired, all holddown arms are released, and the vehicle committed for liftoff. The vehicle rises nearly vertically from the launch pad, for approximately 450 feet, to clear the launch umbilical tower. During liftoff, a yaw maneuver is executed to provide tower clearance in the event of adverse wind conditions or deviations from nominal flight. After clearing the tower, a tilt and roll maneuver is initiated to achieve the flight attitude and proper orientation from the selected flight azimuth. The S-IC center engine cutoff occurs at 2 minutes 5.6 seconds after first vehicle motion to limit the vehicle acceleration to a nominal 3.98 G-load. The S-IC outboard engines are cutoff at 2 minutes 31 seconds after first vehicle motion. Following S-IC engines cutoff, ullage rockets are fired to seat S-II stage propellants, the S-IC/S-II stages separate, and retrorockets back the S-IC stage away from the flight vehicle. A time interval of 4.4 seconds elapses between S-IC engines cutoff and the time the S-II engines reach 90 percent operating thrust level. Following the programmed burn of S-II engines, the S-II/S-IVB stages separate and the S-IVB engine places the flight vehicle in an earth parking orbit.

1-246. POST-FLIGHT DATA EVALUATION.

1-247. The post-flight data is evaluated to determine that the S-IC stage engines operated within the specified values during vehicle launch. The engine parameters are reviewed for abnormalities, sudden shifts, oscillations, or performance near the minimum or maximum limits. The engine performance values are then reviewed and compared to the predicted engine values to determine that all engine objectives were satisfactorily met.

1-248. UNSCHEDULED MAINTENANCE FLOW.

1-249. Unscheduled maintenance consists of those operations required in addition to normal engine and hardware processing, to repair damage, replace discrepant components or hardware, perform modifications and EFIRs, decontaminate, re-preserve, repair thermal insulation, or rectify any unsatisfactory condition. The unscheduled maintenance tasks are done at a specified time and at the location designated, during the normal engine flow process. The locations where unscheduled maintenance can be done are Rocketdyne, MAF, MTF, or KSC; depending on the extent of the task, urgency, capabilities of the location, and how schedules are affected. The location established for complete component maintenance, repair, and testing is the CM&R room at MAF. This facility provides component maintenance support for MAF, MTF and KSC. Limited repairs on components can be made in-place on the engine at MAF, MTF or KSC as directed by the customer. The necessary hardware required for supporting engine and component repairs at field locations is stored and maintained at MAF.

2-250. UNSCHEDULED ENGINE REPAIR SERVICING.

2-251. Unscheduled engine repair and servicing consists of various types of repairs and servicing tasks that are done whenever practical to correct any discrepancies that may exist, perform special inspections, and to update the engine configuration. The various repairs and servicing tasks may include such items as: braze and weld repair thrust chamber tubes, remove and replace components, clean contaminated areas, remove corrosion, touch-up of damaged surface finishes, modifications, EFIRs, post-maintenance tests, lubricate, preserve, and replace desiccants.

1-252. COMPONENT REPAIR.

1-253. Uninstalled engine components from MAF, MTF, or KSC that require repair, modification, analysis or testing are processed in the environmentally controlled CM&R room at MAF. Processing engine components in the CM&R room is required to repair a discrepant component from an engine, perform modifications, failure analysis, inspections, recycle testing, or pre-installation testing. After processing in the CM&R room, the components are designated to be installed on an engine, returned to the engine support hardware center as a spare, returned to the manufacturer, or considered as surplus or scrap. Detailed procedures for component maintenance and repair are in R-3896-3.

1-254. SUPPORT HARDWARE.

1-255. Engine hardware required for supporting the activities at MAF, MTF, and KSC is maintained in the Engine Support Hardware Center at MAF. The Michoud facility is the primary hardware supply center, since the majority of engine and component activity takes place at his location. At MTF and KSC a limited inventory of hardware is maintained to make sure of immediate availability of those items frequently used at these locations. Whenever an urgent need arises at either MTF or KSC, and the hardware required is not locally available, the item is expedited to that location directly from MAF or Rocketdyne.

# 6

# MSFC, Boeing and the S-IC stage

The development of the F-1 engine and S-IC first stage of the Saturn V launch vehicle succeeded essentially because of sound program management. As a whole, the Apollo program was perhaps the most complex American engineering project of the 20th century. As such, it required efficient and effective management and excellent communications at every level within NASA and its many contractors. The F-1 engine program was initiated and managed by the Air Force for the first few years of its development, and then in 1958 was transferred to NASA, and later it and the program to develop the S-IC which would be powered by the F-1 engine were both managed by MSFC in Huntsville, Alabama. While Dr. Wernher von Braun was Director of the ABMA's Development Operations Division at the Redstone Arsenal in Huntsville, after ABMA initiated a missile or rocket program this was managed by the Department of Defense. On the creation of MSFC, this management structure served as the basis of the management of the new civilian launch vehicle programs. As von Braun wrote in 1962:

> The nucleus for the George C. Marshall Space Flight Center was the Development Operations Division of the Army Ballistic Missile Agency. The Development Operations Division was a purely technical organization and we depended entirely on Army organization for administration and technical services. When it transferred to NASA, we had to add administrative and technical service personnel to form a self-sustaining organization. We had no finance office to make out the payroll, no personnel office to hire payroll clerks, no procurement and contracting personnel, and no facilities engineering or support services. Through contractual arrangement we still obtained sizeable support from our Army neighbors, particularly in the utilities area.

Upon its official establishment on July 1, 1960, Dr. von Braun became Director of MSFC. As such, he reported to Maj. Gen. Donald Ostrander, Director of Launch

6-1 Dr. Wernher von Braun (center) with key Marshall Space Flight Center directors at an early meeting on the Saturn launch vehicle. From left: Werner Kuers, Manufacturing Engineering Division; Dr. Walter Haeussermann, Astrionics Division; Dr. William Mrazek, Propulsion and Vehicle Engineering Division; Dr. von Braun; Dieter Grau, Quality Assurance Division; Dr. Oswald Lange, Saturn Systems Office; and Erich Neubert, Associate Deputy Director, Research and Development. (NASA/MSFC)

Vehicle Operations for NASA. Dr. Eberhard Rees became Deputy for Research and Development at MSFC. Delmar M. Morris was recruited from the Atomic Energy Commission as Deputy for Administration. MSFC adopted and adapted the Army's arsenal system of rocket and missile research, development, testing, validation, program management, checkout and launch. In particular, it carried over the central concept of the arsenal system, in the form of various laboratories. Dr. E.D. Geissler directed the Aeroballistics Laboratory. Dr. Helmut Hoelzer ran the Computations Laboratory. Dr. Kurt Debus was head of the Missile Firing Laboratory. Dr. Walter Haeussermann led the Guidance and Control Laboratory. Dr. Ernst Stuhlinger was head of the Research Projects Laboratory. William Mrazek directed the Structures and Mechanics Laboratory. Erich Neubert had the Systems Analysis and Reliability Laboratory. Hans Hueter directed the Systems Support Equipment Laboratory, and also ran the Centaur and Agena upper stage programs. And Karl Heimburg ran the Test Laboratory. In addition, MSFC had specific project offices, each with its own director. These included the Future Projects Office of Heinz-Hermann Koelle, the Saturn Systems Office directed by Dr. Oswald H. Lange with Konrad Dannenberg as his deputy, the Technical Program Coordination Office run by George Constan, the Technical Services Office of David Newby, and others involved with personnel and security. The largest program at MSFC in 1960 was the Saturn I launch vehicle. It was managed by the Saturn Systems Office, and had three project offices run by the

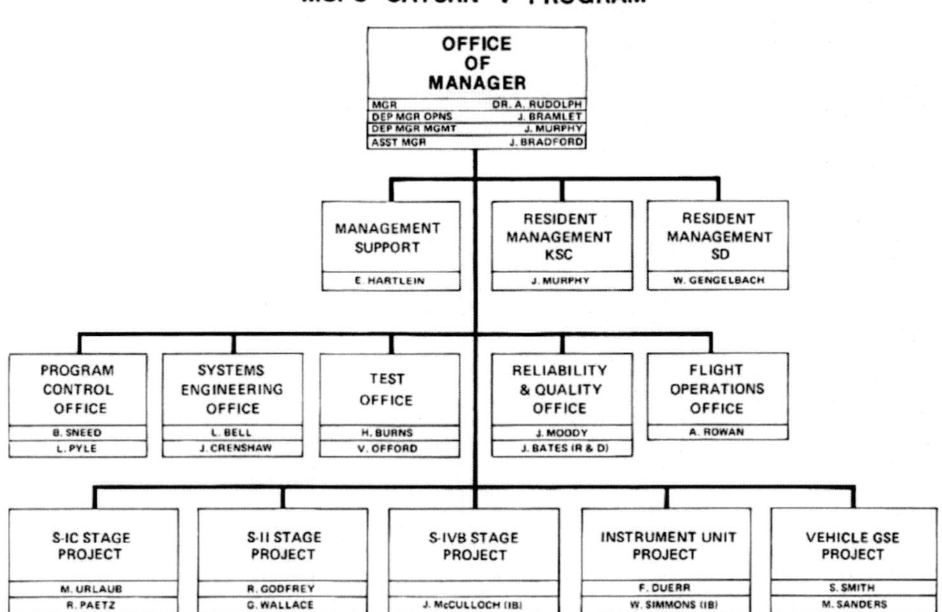


6-2 The success of the Saturn V program rested on the organizational structure and management personnel. Matt Urlaub ran the S-IC Stage Program Office. (NASA/MSFC)

Vehicle Project Manager, the S-I Stage Project Manager, and the S-IV and S-V Stage Project Manager.[1] Dr. Lange created working groups to pinpoint development problems with the major systems of the Saturn I, recommend solutions, implement the recommendations, report on problem resolution, and make the Saturn Systems Office aware of any other issues relating to the launch vehicle. Each working group was directed by a key individual from one of the pertinent laboratories. These early working groups were the first steps towards launch vehicle program management at MSFC which helped to pave the way for managing the Saturn V program in years to come.

## EVOLVING NASA'S MANAGEMENT STRUCTURES

NASA itself underwent a crucial management organizational change designed not only to enable it to manage proposed space exploration programs, both manned and unmanned, but also to establish a structure that would enable it to interact efficiently with its field centers—which, whilst operating autonomously, had to coordinate with Headquarters in implementing its programs. The agency drew on industry executive and military personnel to assist in organizing and managing its programs and their

[1] The S-V stage was eventually dropped from the Saturn I program.

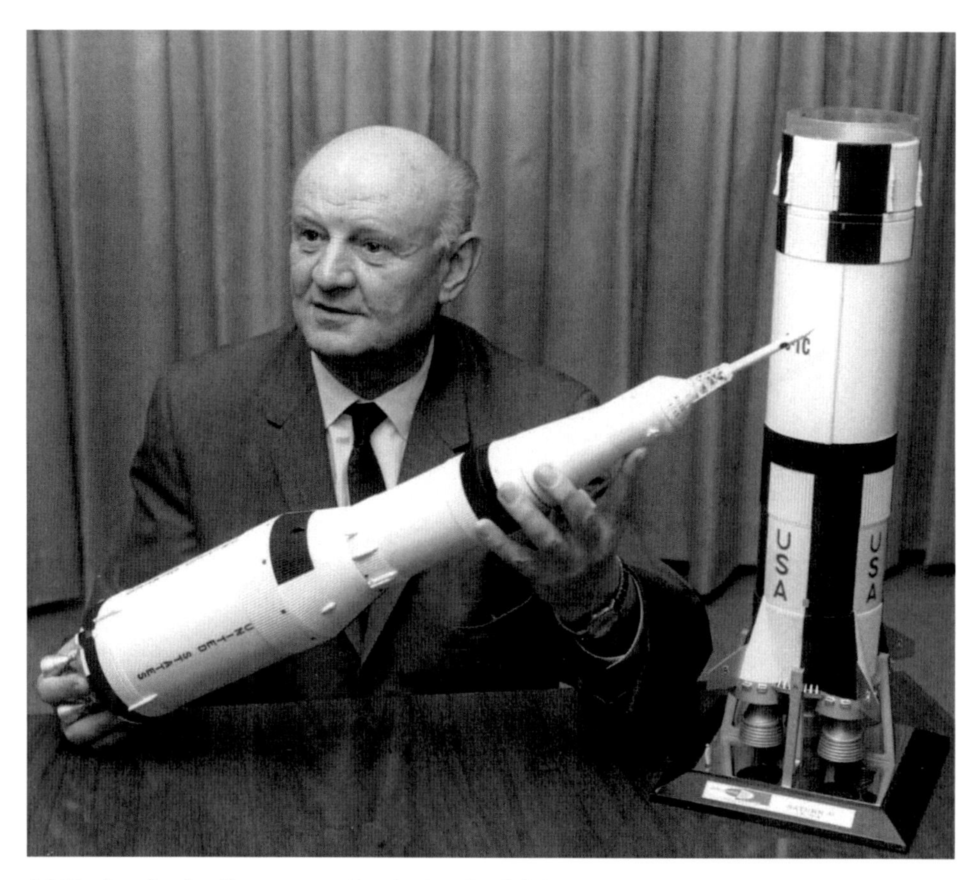


6-3 During the Apollo program Dr. Arthur Rudolph was Manager of the Saturn V Program Office at the Marshall Space Flight Center. (NASA/MSFC)

budgets. Hiring qualified executives was particularly challenging, as this invariably entailed a considerable drop in salary. What could possibly be the incentive to work for NASA? Perhaps, the greatest draw was the lure of involvement in the program to explore the Moon, which combined the most formidable engineering and program management challenges yet attempted with the most dramatic feat of exploration. To many, it was essentially a call to duty. Certainly, the German rocket team members were well aware that they could earn far more working in industry, but they too were driven by the possibility of landing a man on the Moon in their lifetimes. Especially for those at the upper levels of its organization, working for NASA meant a personal financial sacrifice. Fortunately, the agency succeeded in drawing many thousands of skilled and talented people to organize and direct its manned space programs, and in particular the launch vehicle programs.

T. Keith Glennan became NASA's first Administrator upon its founding in 1958. In his three years in that post, and with the assistance of the Space Task Group at the Langley Research Center in Hampton, Virginia, he was involved in developing the agency's organization. Dr. Robert R. Gilruth was Director of the Space Task Group, and relocated with it to the Manned Spacecraft Center in Houston, Texas. Hugh L. Dryden was Glennan's deputy. Robert C. Seamans became Associate Administrator in 1960, and took over as Deputy Administrator on Dryden's death in 1965. George M. Low was on the planning committee for the creation of NASA, and shortly after the agency's formation and his transfer to NASA Headquarters in Washington, D.C. he became Chief of Manned Space Flight. On becoming President, John F. Kennedy appointed James E. Webb to replace Glennan. In September 1961 Webb restructured the agency, creating four new program offices whose directors, along with those of the various field centers, would report to the Associate Administrator. The Office of Advanced Research and Technology was led by Ira H. Abbott. The Office of Space Sciences was directed by Homer E. Newell. The Office of Manned Space Flight was led by D. Brainerd Holmes. The director of the Office of Applications had yet to be named. In addition, reflecting the growing responsibilities of the Space Task Group, Gilruth was named Director of the Manned Spacecraft Center, which was then under construction in Houston.

As James Webb explained in a 1985 interview regarding the NASA management structure he had inherited as the new Administrator:

> The legacy was good, Glennan and his associates had put in place a well conceived foundation and begun the [establishment of elements] of the upper structure on sound theoretical and practical policies. He built for his time and within the framework which President Eisenhower and his administration set; with the goals set by President Kennedy substantially enlarging the effort, we had a different job, but he left us a well-constructed foundation. We built our program on this well-conceived foundation.

On December 21, 1961, D. Brainerd Holmes announced the formation of the Manned Space Flight Management Council, which was to meet on a monthly basis. Its membership included Dr. Robert R. Gilruth from the MSC, Walter C. Williams, Dr. Wernher von Braun and Dr. Eberhard Rees from MSFC and, from Headquarters, George M. Low, Milton W. Rosen, Charles H. Roadman, William E. Lilly and Dr. Joseph F. Shea, who was the Deputy Director of Systems Engineering. Dr. Arthur Rudolph was appointed Assistant Director of Systems Engineering and would liaise between MSFC and MSC on launch vehicle development.

MSFC itself underwent further refinement of its management structure in 1962. In February, the Central Planning Office was formed to assist in planning, coordinating and reporting programs. In the process, some sub-offices were eliminated, new ones created and others realigned. As von Braun wrote in 1962:

> Since the management and direction of the various projects assigned to Marshall constitute our primary mission, we have made two important changes in our management structure. One is the establishment of a central planning office. Our main objectives in establishing this office were to assist top management by providing consolidated overall planning, closer program coordination, and

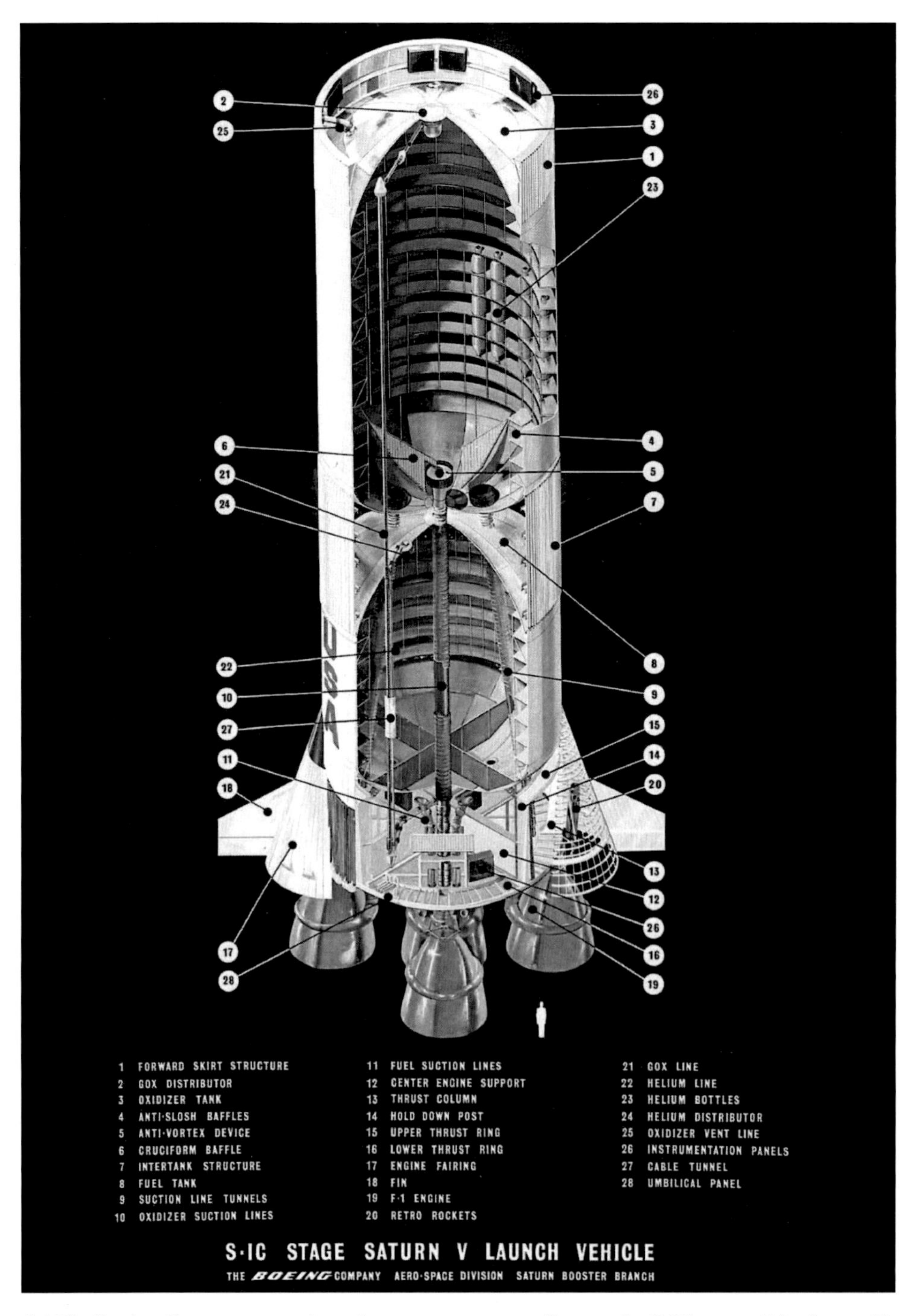


6-4 The Boeing Company was given the contract to manufacture the S-IC stage of the Saturn V, which included integration of the Rocketdyne F-1 engines into the stage. (NASA/MSFC)

6-5 The full scale engineering mockup of the S-IC stage thrust structure assembly nears completion in the Manufacturing Engineering Laboratory of the Marshall Space Flight Center in October 1963. (NASA/MSFC)

6-6 This view of the S-IC stage thrust structure early in its assembly shows the four tapered hold-down posts. (NASA/MAF)

> increased management data and support. Our central planning office is also coordinating the introduction and development of such management tools as PERT (program evaluation and review techniques) and ADP (automatic data processing).
>
> The other important modification, is the creation of project offices, in addition to our technical divisions. The responsibilities of Marshall's ten technical divisions are essentially unchanged from those they possessed in the past. These divisions are not project oriented, but are aligned along professional disciplines such as mechanical engineering, electronics, flight mechanics, vehicle dynamics, and the like.

Two of the distinguishing characteristics of MSFC responsible for its strength and technical competence were 'automatic responsibility' and 'dirty hands engineering.' Automatic responsibility meant that each laboratory was required to solve problems that arose, and if necessary work with other labs to ensure a resolution, rather than simply allowing the issue to go before formal review to determine how it might best be solved. As von Braun put it:

> The division director is responsible directly and solely to me for his performance in each of his assigned disciplines. He is expected to participate automatically in all projects that involve his discipline and to carry his work through to conclusion.

Dirty hands engineering was part of the arsenal culture in Huntsville. It called on program managers to work one on one with MSFC's other offices and laboratories to answer specific manufacturing questions or indeed any other issue. This approach not only expedited resolution of questions or problems and completion of the task at hand, it also expanded an individual's knowledge base. As von Braun explained:

> An important responsibility of the division director is that of assuring the maintenance of a high level of technical competence with his organization. This may be done partially by theoretical studies and by following work being done by other agencies and industry. But [...] it cannot be done adequately unless the technical people of the division keep their hands dirty and actively perform work on projects selected specifically to update their knowledge and increase their competence.
>
> At Marshall, we can still carry an idea for a space launch vehicle and its guidance system from the concept through the entire development cycle of design,

6-7 The S-IC stage thrust structure assembly was the heaviest and strongest section of the first stage of the Saturn V launch vehicle. This one is at the Michoud Assembly Facility. (NASA/MAF)

> development, fabrication and static testing; and we have every intention to preserve and nurture a limited in-house capability. The reason is simple: the ability of American industry to produce fine products is unquestionable. To market them, it has developed a persuasiveness in salesmanship that is unequaled. It's not easy for the Government to determine which bid or proposal we receive from industry is best, and how well competing claims and estimates can be substantiated. In order for us to use the very best judgment possible in spending the taxpayer's money intelligently, we just have to do a certain amount of this research and development work ourselves. We just have to keep our own hands dirty to command the professional respect of the contractor personnel engaged with actual design, shop and testing work. Otherwise, our own ability to establish standards and to evaluate the proposals—and later the performance—of contractors would not be up to par.

All this would prove absolutely vital for Apollo. Massive, complex and powerful machines had to work to near-perfection for the safety of the crews and the success of their missions. The program management structure at NASA Headquarters and its field centers had to be structured in such a way, and function to such a degree, that the formidable goal of landing men on the Moon and returning them safely to earth could be realized within the specified deadline of the end of the decade of the 1960s. To achieve this, Apollo needed the finest organizational and management minds in the country. One such key man was Dr. George E. Mueller, who left his post as Vice President for Research and Development at the Space Technology Laboratories in order to become NASA's Deputy Associate Administrator for Manned Space Flight in the summer of 1963, and upon the departure of D. Brainerd Holmes in September Mueller took over as Associate Administrator for Manned Space Flight.

As Robert C. Seamans recounted in a November 1998 interview with Michelle Kelly for the Johnson Space Center Oral History Project:

> So George came that summer and he did a remarkable job. Brainerd did a great job too. But what George added, among other things, were two things [...] that were really important right at the start. One was to get more senior people running the program. We had a hard time hiring people from industry to come in and take jobs in NASA. Obviously they were going to take a big reduction in pay, and [there were] other reasons it wasn't attractive. We did get some people to come.
>
> The other thing he had was this idea of all up system testing. By then we had fired four Saturn I's. It was just the first stage. The upper stages [sic] had nothing but sand in them. This was the Germanic way of testing, to have lots of vehicles and you take the thing step by step, adding a little bit each time. Before they got a successful V-2, they'd fired seventy-seven V-2s, for example. Even before George came in, it was obvious to me that we weren't going to be able to land on the moon in the decade or even come close to it if we kept proceeding in the same sort of plodding way. But George came in and he said, 'All up system testing.' He said, 'The very first Apollo launching will be with the complete vehicle, everything.'

6-8 A Boeing technician at the base of a S-IC stage thrust structure. (NASA/MAF)

The Huntsville people in particular were absolutely aghast at that. They said, 'It'll never work.'

Mueller was both brilliant and tireless. He often worked seven days a week, and expected others to do so—often holding meetings on Sundays. He recognized the absolute necessity of getting experienced men to come to NASA to manage Apollo and the many programs which were part of it. He used his contacts in industry and the Air Force to persuade people to join the team. Mueller contacted Gen. Bernard A. Shriever, who was highly regarded for his successful management of the ICBM program, and sought the services of a number of people. One, Brig. Gen. Samuel C. Phillips, had been involved in the successful B-52 bomber program and then headed the Minuteman ICBM program. On December 31, 1963, NASA announced that he had been appointed Deputy Director of the Apollo Program Office at Headquarters. In the meantime, on November 22, tragedy struck the nation with the assassination of President John F. Kennedy. America's most vocal champion of lunar exploration, and the one who had set the program in motion, was now dead. Even as the nation mourned and millions of American's dealt with the shock, those within NASA and

6-9 The lineup of the F-1 Engine Checkout Stations at the Michoud Assembly Facility. (NASA/MAF)

the Congress were resolved to push on with Kennedy's audacious challenge. The facilities for preparing and launching missions for this program at Cape Canaveral in Florida were renamed the Kennedy Space Center in his memory.

As Dr. Mueller stated during a 1999 interview for the Johnson Space Center Oral History Project:

> General Phillips was probably the best manager that I've ever known. I'd known him when I was at TRW (Thompson-Ramo-Wooldrige), or STL (Space Technologies Laboratory), and had known of his work, furthermore, on the B-52 bomber. He had a long career of very outstanding management responsibilities and successes. So when I went to NASA, one of the first things I did was to seek support from the Air Force to bring in some people who were knowledgeable about program management, because NASA never really had the opportunity to manage programs of any size before. So they didn't have the infrastructure to be able to carry it out—to carry out a program like Apollo. That was the turning point, really, in the program itself, because until we had a sufficient infrastructure to be able to put in place the kind of management practices that are essential for a system-wide approach, we were having trouble. So, the combination of that plus the establishment of autonomous program offices reporting back to Sam Phillips in parallel with reporting of the center directors back to me gave us the kind of insight and controls we needed to carry out the program.

Another crucial decision which Mueller made in the latter part of 1963 was the establishment of the NASA-Industry Apollo Executives Group. This included the program managers within the Office of Manned Space Flight and top executives—often CEOs—of the companies that had prime contracts with NASA. Meetings were held monthly in Washington D.C., with Dr. Mueller and those executives, or their deputies, expected to attend without fail.

In 1963, MSFC underwent a significant reorganization, with the introduction of two new divisions. Herman Weidner, a veteran of von Braun's team, was appointed as Director of Research and Development Operations, and Robert Young was hired from Aerojet as Director of Industrial Operations. As part of the restructuring, all the center's laboratories were placed under the umbrella of Research and Development Operations, and the two Saturn programs and their engine programs became part of Industrial Operations. The Saturn Systems Office was dissolved, and replaced by the Saturn I Office, the Saturn V Office, and the Engines Office. Konrad Dannenberg became Director of the Saturn I Office, Dr. Arthur Rudolph was transferred from Headquarters as Director of the Saturn V Office, and Leland F. Belew was assigned as Director of the Engine Project Office. The fact that Belew had been head of the former Engine Management Office within the Propulsion and Vehicle Engineering Division made this a smooth transition. The Engine Project Office was responsible for managing the H-1, RL-10, J-2, and F-1 engine programs. This particular year also saw personnel additions at MSFC that profoundly influenced the direction and management of the Saturn V launch vehicle. One such appointment was Bill Sneed, who became Director of the Saturn V Program Control Office. As he recalled to this author in 2006:

> In 1963, Dr. Arthur Rudolph, I and several other people joined the Marshall Space Flight Center. I had been with Dr. Rudolph on the Pershing missile program. When NASA was formed, Dr. von Braun and his team transferred to Marshall Space Flight Center. Those of us in the Pershing program office remained with the Army to complete the development of the Pershing. I had responsibility for the budgeting of an element within that program.
>
> We were initially assigned to Dr. Joseph Shea at NASA Headquarters, who was in charge of systems engineering and integration for the Apollo program. He had resident offices at the Manned Spacecraft Center in Houston, Texas, at the Marshall Space Flight Center in Huntsville, and at the Kennedy Space Center in Florida. Up until 1963, Dr. Oswald Lange had been head of the Saturn program. Dr. Rudolph assumed responsibility for the Saturn V program and I joined the program at the same time. I had responsibility for developing the schedules, for monitoring progress and problems associated with meeting those schedules, and recommending corrective actions for those. The second major function was that of the budget. I had responsibility for the overall Saturn V budget planning, and the receipt distribution and control of all the funds for the Saturn V program.

The Saturn V was an enormous engineering and program management challenge. Dr. Rudolph streamlined the Saturn V Program Office and entrusted its operation to

6-10 In the early phases of the Saturn V program, the S-IC stage was assembled in the Horizontal Assembly Building. Here we see the thrust structure and RP-1 tank assembly being mated to the interstage adapter. (NASA/MAF)

several key individuals. Among them was Jim Murphy, who was Resident Manager at KSC for the Saturn V.

As Konrad Dannenberg recalled to this author:

> Arthur Rudolph put up a Saturn V management office, and it was admired by many people in private industry. Many people came to Huntsville to see his management system for the Saturn V program. Rudolph picked as his deputy, James Murphy, which was another clever decision he made. Murphy was an American Air Force officer. This was a very meaningful decision. Of course, on the other hand, von Braun relied quite a bit on Arthur Rudolph. In the beginning of the Saturn V program, we had daily contact with him because our offices were right next to each other. Von Braun had, as a basic routine, a Monday morning meeting that included not only the key people involved in the Saturn program, but also the other programs. We had to give a brief report, and also to hand in a written report. He apparently went into quite some detail through all these written reports, made notes on them, and sent them back to those who wrote them.

6-11 The lower section of an S-IC stage being moved at the Michoud Assembly Facility. This is a good view of the thrust structure. Note the five mounting blocks to accept the mating portion of the F-1 engine gimbal assembly. (NASA/MAF)

Dr. Rudolph's successful management of the Pershing program gave him proven methods of how to structure the Saturn V Program Office, and how to manage the program itself. He also worked with others on the Pershing program in whom he had total confidence to manage their respective areas, and asked some of them to join the Saturn V program. Rudolph's views of program management aligned with the newly reorganized MSFC, in particular regarding 'automatic responsibility.' As he said in an interview conducted in 1968:

> There are two categories of people I have dealt with as a manager. One category, you only have to touch on the problem and if the guy has been with you long enough he will catch on right away. And without long discussions. He knows what the problem is. He knows that this is what bothers me and something needs to be done, and if he says yes I'll do it I didn't have to worry about it. It would be done. These are the great guys. The guys who really can do what you had in mind and carry it through. They, in turn, of course, have to convey the message to others, and it becomes more and more difficult the further down [in the organization] you go. The other category, they hear it and they have the best intention but they lack the capability to stay behind it and see that it is being done. I think I was fortunate that most of the fellows who worked for me—my product managers, [Saturn]

stage managers or hardware managers, and my functional managers—in the majority of the time would go and do it and I wouldn't have to check. I talked about the hardware managers, or stage managers or product managers which is the best generally accepted term. That is one category of manager. If you look at a usual organization chart, you call that the line [organization] even if you are in a program office. Then you have sort of a split past there—the staff. I made it clear to my fellows that they were not staff. They worked for me as managers just as the hardware managers. You have the responsibility, and authority, and *you manage*. You function (like quality and reliability). And therefore, my whole setup was a matrix. The horizontal cut was the hardware, and the vertical cut was the functions. Engineering goes across the board.

Systems engineering is most important because it is the least understood. There is a lot of confusion through the world. And in other engineering functions, quality and reliability, testing—it's an engineering function but you don't call it that because otherwise it gets more confusing and you can't handle it, manage it. It is too much for one guy. You break it up. Therefore, you have the verticals: systems engineering, testing, quality and reliability. Then, you take the whole stuff—the hardware—the functions—all managers, not staff—and put them in an envelope—that's program control. That is what is needed. This is my particular concept. Many, many people don't agree with it, but it worked.

How was the Saturn V program structured so that its three stages, instrument unit, Lunar Module and Lunar Module Adapter, Service Module, Command Module and Launch Escape System could be efficiently integrated? Here is where Dr. Rudolph's engineering and management experience influenced the pace and success of the Saturn V program. His Saturn V Program Office had to interface with other NASA centers to ensure that all the concurrent engineering of the Saturn V elements, their manufacture, testing, flight readiness and delivery to KSC flowed smoothly. MSFC managed the programs for the S-IC, S-II, and S-IVB stages, including their engines, the Instrument Unit, and issues relating to the spacecraft. MSC in Houston was responsible for managing the programs for the Lunar Module and Lunar Module Adapter, Service Module, Command Module and its Launch Escape System, and all related engines, thrusters and support equipment.

There were five key offices within the Saturn V program at MSFC. These were the Program Control Office directed by Bill Sneed, the Systems Engineering Office, directed by L. Bell, the Test Office directed by H. Burns, the Reliability and Quality Office directed by J. Moody and the Flight Operations Office directed by A. Rowan. There were two office buildings behind the main headquarters block, Nos. 4201 and 4202, and the Saturn offices and Saturn engine offices were located in No. 4201. Management of the Saturn V was split into the S-IC Stage Project, the S-II Stage Project, the S-IVB Project, the Instrument Unit Project, and the project which dealt with the spacecraft.[2]

[2] In NASA parlance of the time, the integrated launch vehicle and spacecraft was referred to as the space vehicle.

6-12 A close-up of one of the S-IC support blocks located beneath the hold-down post. These four blocks supported the entire weight of the Saturn V on the Mobile Launch Platform hold-down arms. The hold-down arms did not release until the five F-1 engines had achieved full and stable thrust. (Anthony Young)

Bill Sneed explained to this author in 2006 how the management of these projects was handled at MSFC:

> We had a fairly unique organizational structure [that] we called mirror-image offices. Kennedy Space Center, Marshall Space Flight Center and Johnson Space Center had identical organizational structures. That was also true of the Apollo Program Office in Washington. For example, there was a program control office in headquarters, and there was one in each of the centers. On a day-to-day basis, program control people were in constant communication with one another on how things were progressing, which kind of problems we were having schedule-wise or budget-wise, or technically. If actions had to be taken, we would work that through the program and project managers to take corrective actions. With these communications, we had our finger on the pulse of what was going on with our counterparts.
>
> We used a very key tool, and that was a work breakdown structure. You have to break a big job down into smaller jobs. The work breakdown structure was a

two-tiered structure. The hardware elements were comprised of the S-IC stage, the S-II stage, the S-IVB stage, the instrument unit, spacecraft, F-1 engines and the J-2 engines. And then, we had another tier, and these elements were program control, system engineering and integration, system test for reliability and quality assurance.

For the hardware elements, which were 90 to 95 percent of the total budget, we had technical requirements, schedules and budgets established and controlled for each of those elements. We had managers for each of those hardware elements, and each of those elements had a program control which was responsible for that element. We worked through those managers in identifying the budgetary needs, technical requirements and how well they were meeting them, and schedules. If one stage was making more progress than another, we had to balance out the resources. My job was to make recommendations on how to restore that balance. We also maintained a rather hefty program reserve, which Dr. Rudolph was responsible for and which I managed. If one of the project elements was in trouble, we had the means of using that reserve to buy back the time if we were running behind schedule or if we encountered particular problems. The big job was trying to keep everything moving along at a pace so all the elements came together at the same time.

Monthly meetings were held in the MSFC Program Control Center to discuss the progress or problems of all the elements of the Saturn V program. Before the age of computers, PowerPoint and other presentation means, the Control Center used large, clear Plexiglas panels with hand-taped schedules that were back-lit for emphasis and clarity. At a glance, all the managers and engineers present could see the status of each element of the Saturn V.

As Sneed explained:

We had schedules up on the Plexiglas panels for the F-1 engine, the J-2 engine, the S-IC stage, S-II stage, S-IVB stage—every hardware element of the program. We also had schedules in there for every vehicle, such as the first flight article, second flight article, and for all of the ground test articles. We had schedules for all the dynamic and all-systems tests. We had all the things required to conduct that test, displayed on those Plexiglas panels, and these were maintained at all times. There was a name on each chart stating who had responsibility for maintaining that on a real-time basis, keeping it up to date. If there was an issue, for example if an engine delivery was late or if the S-IC stage was late on delivery, we would install a red arrow on the schedule to identify that as a potential problem.

We also had a problem resolution board that identified what we considered to be our top ten problems in the program at any time. I had knowledge of everything that was going on in the center, and then I would be make sure that on the agenda we'd address every major problem that had been identified—be it technical, schedule or budget. We tracked the resolution of each problem, and as you can imagine, those subjects were always covered in our program review until the problem was resolved. In special meetings where we were dealing with a particular

6-13 An S-IC stage is prepared for installation of the five F-1 engines. (NASA)

> problem, we would meet with the appropriate people, and many times Dr. von Braun would be involved with those to brief him on a particular problem—for example, [we had] stages blow up on the test stand,[3] and that was a major problem!

Dr. Rudolph proved to be exactly the right man at the right time to manage the Saturn V program at MSFC. The men he surrounded himself with often marveled at his ability to properly determine the cause of a particular problem and prescribe its solution—whether it involved manufacturing, scheduling, quality control, testing or budgeting. He was, in fact, the second most experienced and powerful man at MSFC after Dr. von Braun himself. The two men shared many management traits.

Speaking of his years of working with Dr. Rudolph, Sneed said:

> He was just a tremendous person to work for. I told him once, 'It seems you have a sixth sense in your decision-making process.' I was always impressed with how he approached problems [and it] finally dawned on me that the sixth sense I was

[3] On May 28, 1966 the S-II-T stage hydrogen tank exploded on a test stand due to the tank pressure sensors having been disconnected.

> seeing was that of experience. He brought a tremendous amount of experience to the program.

Dr. Von Braun's management style and interpersonal skills were of the highest order, and he was given the utmost respect from the highest level managers down to the maintenance crews. He was gregarious but courteous, and often enjoyed being in the media limelight. Dr. Rudolph shunned the media attention—indeed, he was often an unknown face to the media. He was all business. He hand-picked his team, knew each individual's abilities, and assigned them the maximum responsibility to match. In return he expected results, and allowed no excuses. He truly embraced 'automatic responsibility.'

Bill Sneed worked closely with both Dr. von Braun and Dr. Rudolph, and noted their different management styles:

> I would have to say Dr. von Braun was at the top of my list of anyone I dealt with at Marshall Space Flight Center. In terms of his knowledge, he was a true leader. He was a compassionate man. He dealt with people gently but he could pound on the table when he had to, and he did on a couple of occasions; but he was real gentleman. I did not fear going before him at all, but I did have to be prepared. Dr. Rudolph—unlike Dr. von Braun—was a low-key guy. He often referred to himself as a blacksmith. Von Braun was a suave and charismatic man who could speak with presidents—and he often did. Dr. Rudolph was an entirely different personality. He was the kind of man who got his satisfaction by working in his office, working to get the job done. He was not one to go to the press and conduct a lot of public interviews. That wasn't his style.

## THE S-IC STAGE IS CHOSEN FOR THE SATURN

During 1961, MSFC had prepared the Request for Proposal for the first stage of the Saturn C-5, as the launch vehicle that would become the Saturn V was designated at that time. On September 17, NASA invited 36 companies to bid for the contract to develop the S-IC stage powered by a cluster of F-1 engines. A bidders briefing was held in New Orleans, Louisiana on September 26, where the details of the RFP were revealed. On November 10, MSFC received five bids. Over the next month it held discussions and negotiations with the potential contractors. On December 15, 1961, MSFC issued a press release saying that the Boeing Company had been selected as a possible prime contractor. The word "possible" was essential, since negotiations on a range of issues about the stage's design, fabrication, testing, and delivery of flight stages were still under negotiation; but just as significant was the fact a decision had not yet been made between the Saturn C-5 and the Nova as the launch vehicle to be built to send a manned mission to the Moon.

6-14 Installation of the F-1 engines begins on an S-IC stage at MAF. (NASA)

## Four or five F-1 engines?

Matt Urlaub joined MSFC shortly after Boeing was chosen as the provisional contractor for the S-IC. He had worked at the Redstone Arsenal in the 1950s on various missile programs, including the Redstone and the Jupiter, often working on-site at contractor facilities. His ability and character were known to Dr. Rudolph and Dr. von Braun, and he had distinguished himself while in the Army. He was assigned as the S-IC Project Manager, reporting to Dr. Rudolph.

Historian Roger Bilstein interviewed Urlaub about the decision to add a fifth F-1 engine to the S-IC stage, but his comments were not quoted in *Stages to Saturn*. As Urlaub explained:

> As I remember the conversation or deliberations on this point, when we were in the four-engine configuration, that which we bid on, we were considering performing the mission by putting two Saturn Vs in Earth orbit in assembling the command module, service module, and I think final propulsion stage for escape velocity—doing all that in Earth orbit and then going on to the Moon from an

Earth parking orbit. Now with four engines you did not have any other option. You didn't have the option to go direct to the Moon and do your lunar orbit maneuvers where you separate and part of you stays in lunar orbit, and the lunar module goes down to the Moon and then back up to rendezvous. You didn't have that option with the four. You have that option with the five, and also the advantage that you do it with one vehicle.

[On the one hand we had] the safety of an Earth orbit assembly operation where the men are in view, they have options to return to Earth if things don't go right, but it's more costly, you need two vehicles to do it, high launch rate. That set a trade versus [. . .] simplified operations on the ground for one launch, and then assume the added risks of separation and rendezvous in lunar orbit where the men don't have the option to return safely to Earth if something goes wrong. That's always a very tricky maneuver, and it happens behind the Moon where we're out of transmission with them. They have to fire the service propulsion module to go into a lunar trajectory for landing, and when they fire to come home they have to fire the service propulsion module behind the Moon. It's strictly an astronaut function. There is no protection of the control center monitoring their operations, making sure they're going through their check list correctly. It's strictly up to them. They are on their own.

It's those kinds of trades that were deliberated and the decision was finally, yes—we'll go to five engines. We'll have the vehicle that has the capability to go directly to the Moon with everything on board that's necessary to perform the mission. One of the things that I think swayed that decision, which came out a little later in the deliberating process, was this concept of 'free return,' where if something goes wrong during powered flight or in that we call the translunar trajectory, as happened on Apollo 13, [then] if they do nothing the trajectory is such that she goes around the Moon and comes back toward Earth [although] in the return trajectory they do have to do some correction to bring it back in. But that free return, I think, was the thing that helped make the decision to go to a single launch capability—which required five engines.

On December 21, 1961, D. Brainerd Holmes, Director of the Office of Manned Space Flight, announced the establishment of the Manned Space Flight Management Council. Its members included Robert R. Gilruth, Walter C. Williams, Wernher von Braun, Erberhard Rees, George M. Low, Milton W. Rosen, Charles H. Roadman, William Lilly, Joseph F. Shea and D. Brainerd Holmes. These men were among the most powerful in NASA. This council would meet once a month to formulate plans and resolve problems regarding the manned space flight programs. At the Council's first meeting (which was on December 21) it was decided to add a fifth F-1 engine to the S-IC stage, that Boeing should be the contractor for stage integration of the Saturn C-5, and that the company should proceed with a detailed design study and development and production plan prior to NASA letting final contracts. The decision to add the fifth engine to the S-IC made the discussions about possible lunar orbit rendezvous much more positive. On February 14, 1962, NASA and Boeing signed the contract for the planning and research phase of the S-IC stage, in expectation of later

6-15 The vast Michoud Assembly Facility near New Orleans, Louisiana. The tallest building is the S-IC Vertical Assembly Building. (NASA)

receiving the contract for the design development, manufacture, test and launch operations. Over the next four months, NASA and its launch vehicle and spacecraft contractors debated the best mode for mounting Apollo lunar missions. On June 22, 1962 the Manned Space Flight Management Council made the momentous decision to use the Saturn C-5 launch vehicle and lunar orbit rendezvous. On July 11 this was announced publicly. In February 1963, the name of the vehicle was simplified from Saturn C-5 to Saturn V, using the Roman numeral for five. On February 25, NASA signed the formal contract with Boeing for design, development and manufacture of the S-IC stage, valued at $418,820,967.00. The contract called for one ground test stage and ten flight stages.

## S-IC stage contract negotiations

Drawing on the extensive contacts developed whilst at ABMA, Matt Urlaub, head of the S-IC Stage Project Office, brought in experienced people to work on the project. One was Bill Hallisey. He had moved from ABMA to work in MSFC's Propulsion and Vehicle Engineering Laboratory, and later in the Vehicle Integration Group. As Hallisey recalled to this author in 2007:

> When Boeing was brought on board and [...] given the complete design and manufacturing responsibilities for the first stage, they went through the whole

design and revalidated the Marshall concept for their own purposes because they were taking responsibility for the stage. It was the philosophy of MSFC that even although the contractor had full responsibility, MSFC was going to perform a major oversight of what was going on and delve deeply in the program operation.

The initial contract, where the selection was made, was done while I was still at ABMA. After I came to Marshall, we did a lot of contract modification work. We switched the type of contract from a cost plus fixed fee contract to a cost plus incentive fee contract. And that in itself was a major undertaking because it was sort of the first of its kind within the agency. We finally came to an agreement with Boeing as to contract conditions and who was going to be responsible for what—getting that line of responsibility clearly defined. It was a delicate set of negotiations that we went through to make sure both parties were in full agreement with the final result.

## Model testing and S-IC base heating from the F-1 engines

As it had with the Saturn I and Saturn IB launch vehicles, MSFC developed a model testing plan for the Saturn V, and specifically the S-IC first stage. The purpose of the testing was to gather vital information on the issue of base heating from the cluster of five F-1 engines and diminishing atmospheric pressure from wind tunnel tests that would simulate the first 30 km of powered ascent and trajectory data from 38 km to about 61 km of altitude. The areas of test evaluation with respect to the F-1 engines were base heating, base convection and base radiation, but base pressure and base environment were also investigated. An earlier study had shown that long duration test techniques would not be necessary, and that, based on the experience gleaned from testing models of the S-I and S-IB stages, a short duration technique would be adequate. The 1:45 model was designed by the Cornell Aeronautical Laboratory and fabricated by MSFC. At this scale, the S-IC model had a diameter of 22.5 cm. As the upper stages were not present, the model was fitted with a forebody which had an aerodynamic nose, and this was attached to the pylon of the wind tunnel. The model was fairly detailed in the operation and configuration of the scale F-1 engines, and it had four engine fairings to reduce aerodynamic loads on the model engines—just as the fairings on the actual vehicle would do. The outboard engines could gimbal. A single combustion chamber installed above the engines burned a mixture of ethylene and gaseous oxygen which was fed to the engine nozzles to indicate the flows. The model could even simulate turbine exhaust. Instrumentation to record temperatures and pressures was affixed in numerous places on each engine model, and across the base plate. Four different test facilities were used: the 2.44 × 2.44 meter Transonic Wind Tunnel at Cornell, the 2.44 × 1.83 meter Transonic Wind Tunnel at the Lewis Research Center, the 3.0 × 3.0 meter Supersonic Wind Tunnel at Lewis and the High Altitude Chamber at Cornell.[4]

[4] The name of the Lewis Research Center was changed in 1999 to the Glenn Research Center.

6-16 A sequence of photographs showing the assembly of the S-IC stage in the Vertical Assembly Building at the Michoud Assembly Facility. (NASA)

The tests began in February 1963 and concluded in August 1965. Among the key conclusions were:

1. The addition of deflectors significantly reduced convection heating.
2. Convective heating was a minor contributor to base heating below 30 km.
3. High convection heating rates occurred on the outboard nozzles above 36 km; center engine convection was much less.
4. Data showing a hot base plate established that exhaust gases were present in this region at low altitudes. At high altitudes with lower atmospheric pressure, a significant portion of the exhaust gases were present at the base.

Among other findings that were applied to the design of the S-IC stage, the model test results were used to properly shield the base of the stage with thermal panels, and confirmed the requirement for a thermal protection system for the exterior of the F-1 engines.

## S-IC stage configuration

There were five key segments to the S-IC stage: the tail section that incorporated the thrust structure and F-1 engines, the fuel tank structure, the intertank skirt structure, the LOX tank structure, and the forward skirt structure. The propellant tanks were each domed at both ends, and the cylindrical sections integrated the outer skin of the stage. The length of the entire stage from the top of the forward skirt structure to the trailing edge of the F-1 engines was 42.0 meters, and the diameter of the cylindrical section was 10.0 meters (the fairings and fins at the base projected further out). The LOX tank was installed above the fuel tank. LOX would be delivered to the engines by five tunnels passing through the fuel tank, each exiting at the base of the fuel tank with a prevalve that would direct the LOX to one of the F-1 engines. There were ten fuel suction lines exiting at the base of the fuel tank with prevalves feeding the two fuel inlets of each F-1 engine.

The thrust structure was the heaviest of the five major components, weighing 21.8 metric tons. Its major components were the F-1 engine-support fittings, a lower and an upper thrust-ring assembly, five F-1 engine thrust posts, four hold-down posts and the center engine support beam. Five intermediate circumferential rings and skin panels made up the remainder of the thrust structure. The five thrust posts served as the engine attachment points. The four outer thrust posts were an integral member of the outer thrust structure skin after assembly. The center engine support assembly was the cross structure with the center thrust post on the centerline of the stage. The majority of the thrust structure was created from high-strength 7178 aluminum alloy. The center engine thrust post and the massive hold-down posts were machined from forged 7079 aluminum alloy. The outboard thrust posts measured 0.91 × 1.0 meter at the base, tapered to 0.35 × 0.56 meter at the top, and measured 4.27 meters in length. They were machined from a solid billet of 7075-T6 aluminum, and were the longest-lead-time item in the entire thrust structure. The outboard F-1 engines were mated to the thrust posts using adapter fittings bolted to the engine gimbal blocks.

There were other key structures of the aft section thrust structure related to the F-1 engines. The base heat shield comprised steel honeycomb panels covered by thermal insulation. They were fitted just forward of the gimbal plane of the F-1 engines, to provide protection to all the engine and stage components above them. These panels were permanently fixed, with access panels in each quadrant to provide access to the internal areas of the thrust structure. Flexible thermal curtains were fitted around the propellant line interfaces to plug gaps in the heat shield. The four F-1 engine fairing heat shields had curved thermal panels on the inside surfaces from the trailing edge of the fairing to the base heat shield intersection.

## The S-IC stage transporter

After assembly, the S-IC stage would need to be transported relatively easily to the various test facilities, and finally to KSC. The task of designing the S-IC transporter fell to the MSFC Test Laboratory. It was decided to employ a pair of large dollies to support the ends of the stage. The front dolly had a suitably strong framework with four two-wheeled trucks on each side, with a total of 16 pneumatic tires. Each truck had hydraulic steering and could turn through 90 degrees to allow lateral movement. The rear dolly had two, two-wheeled trucks on each side, with eight pneumatic tires, and could also turn through 90 degrees for lateral movement. The transporter was to

6-17 An assembled S-IC stage is rotated for loading onto the S-IC Transporter inside the Michoud Assembly Facility. (NASA)

6-18 The S-IC-D Dynamic Test Stage was the first S-IC stage to be produced by the Michoud Assembly Facility. It had one F-1 engine and four envelope mockups. (NASA)

6-19 The first two S-IC flight stages were both assembled and tested at MSFC. The third was assembled at the Michoud Assembly Facility and tested at MSFC. All later S-IC stages were manufactured at Michoud and loaded onto barges (shown here) for testing at the Mississippi Test Facility. (NASA)

be moved by a modified Army M-26 tank retriever. In August 1963 MSFC awarded a contract for the construction of two transporters. These served a vital role in the assembly of the first few S-IC stages, which had to be done horizontally prior to the completion of the Vertical Assembly Building at the Michoud Assembly Facility.

### S-IC stage manufacture at MSFC and the Michoud Assembly Facility

With the selection of Boeing as the prime contractor for the S-IC stage, and signing of the production contract in December 1961, the company worked with MSFC on the details of the S-IC stage and the design of fabrication and manufacturing tooling, which was to be made at Boeing's facility in Wichita, Kansas. Boeing established an office in the Huntsville Industrial Center Building. Hundreds of people were needed to support work by Boeing on the massive first stage of the Saturn V. This included designers, drafters, engineers and other personnel. By mid-1962, in addition to 600 people in its own Huntsville office the company had almost 500 people working at MSFC. Once the Michoud Assembly Facility was set up to produce S-IC stages, the company would have over 400 people working there. However, MSFC had a vital role in the construction of two non-flight stages and the first two flight stages. This phase of the S-IC program prior to the production operation starting at Michoud was also key to MSFC's determination that the Saturn V booster would indeed be flight-ready. The synergy between Boeing and MSFC personnel in Huntsville proved to be highly beneficial to the program. Among the first S-IC segments built at MSFC was the tail section mockup, to which F-1 engine mockup and F-1 test engines were mated to verify the overall fit, the alignment of propellant feed lines and cable interfaces and the requisite clearances.

The first S-IC stage made at MSFC was the S-IC-T. It was manufactured during 1964 and early months of 1965 and installed in the S-IC test stand in March 1965, at which time the F-1 engines were installed in the vertical configuration. The second S-IC stage built at MSFC was the S-IC-S, known as the Structural Test Stage. It was also made between 1964 and 1965. Structural and captive firing tests began in April 1965, and a number of its individual components also underwent specific testing. It continued to be used for structural testing through 1967. Assembly of the first flight stage, designated S-IC-1, began in 1964. The thrust structure was built by Boeing at the Michoud Assembly Facility and shipped to MSFC. Both the Vertical Assembly Facility and the Manufacturing Engineering Laboratory's horizontal assembly area were used in the assembly of the first two flight stages. The five F-1 engines were installed horizontally in the Manufacturing Engineering Laboratory over a two week period in August 1965. S-IC-1 was completed by the end of the year and underwent post-manufacturing checkout in January 1966. Assembly of S-IC-2 began at MSFC early in 1965. Manufacturing of the propellant tanks, thrust structure, intertank, forward skirt and assembly of the stage occupied the remainder of the year. The five F-1 engines were fitted to the thrust structure at the rate of one per day, concluding on December 11, but contamination from hydraulic fluid was found in the engines and they had to be removed, cleaned and reinstalled in April 1966.

NASA selected the Michoud Ordnance Plant outside of New Orleans, Louisiana as the Michoud Assembly Facility for the S-IC stage on September 7, 1961. It had been used during World War II to manufacture ordnance, and by Chrysler during the Korean War to manufacture tank engines. Boeing converted it for manufacturing the S-IC stage. The S-I and S-IB stage were also manufactured there. Boeing worked with Rocketdyne personnel and other S-IC subcontractors for the design of the S-IC manufacturing flow and post-manufacturing checkout. The first S-IC manufactured at the Michoud Assembly Facility was S-IC-D, or Dynamic Test Stage. Although a non-flight stage, this was vital for checking the structural integrity of the S-IC under various load and stress conditions. Boeing commenced fabrication and component manufacturing of it in April 1964, but because manufacturing procedures needed to be verified and in some cases changed, its assembly took over a year. The stage was essentially complete by June 1965. One F-1 engine and four F-1 envelope mockups were fitted. It was used in many tests over several years to gain vital information on the stage. It was later donated to the Alabama Space & Rocket Center in Huntsville, and is now part of an impressive indoor Saturn V display. The next non-flight stage manufactured at the Michoud Assembly Facility was the S-IC-F. This was for use as the Facilities Checkout Vehicle. Work began in 1964 and was complete by the later months of 1965. In October 1965 an F-1 mockup engine experienced minor damage from the installer tool used to align and install the F-1 to the S-IC thrust structure.

Changes were made to the installer tool to ensure that production engines could not be similarly damaged. Like S-IC-D, this stage received one F-1 engine and four F-1 envelope mockups. After post-manufacturing checkout in November 1965, the stage was shipped by barge to KSC in January 1966. It was taken to the Vehicle Assembly Building and installed on a Mobile Launch Platform in March 1966. On May 25 the assembled 500-F Saturn V was transported to Launch Complex 39A to demonstrate the mobile launcher concept.

The first S-IC flight stage assembled at Michoud was S-IC-3. Its assembly began in the fall of 1964 and was completed roughly a year later. Work on S-IC-4 began in the spring of 1965, and Boeing established a phased stage assembly schedule for the remaining flight stages. S-IC-12 was the last flight stage for a manned Apollo lunar mission—Apollo 17. S-IC-13 was used to launch Skylab. S-IC-14 was never flown, and is now housed indoors as part of a Saturn V display at the Johnson Space Center in Houston. Although S-IC-15 underwent test firing at the Mississippi Test Facility, it was later returned to Michoud, where it is now on public display as a testament to the manufacturing skill of the NASA and contractor teams who worked to make the Saturn V program a complete success.

# 7

# Testing the F-1 engine and S-IC stage

The Apollo program required NASA to develop engineering, manufacturing, testing and launching facilities on an awesome scale. The civil engineering and construction projects to support the program transformed the industrial landscape of the United States. The facilities were designed not only with Apollo in mind, but also the long term future of America's manned space program, and indeed many of them will be used in support of Project Constellation. Facilities to test the F-1 rocket engine and its components were constructed at Rocketdyne's Santa Susana Field Laboratory in California, the Edwards Field Laboratory at Edwards Air Force Base in California, and MSFC in Huntsville, Alabama. The S-IC stage of the Saturn V launch vehicle contained a cluster of five F-1 engines. Stage development and acceptance testing of flight stages S-IC-1 and S-IC-2 was done at MSFC and at NASA's Mississippi Test Operations—later renamed the Mississippi Test Facility and, later still, the National Space Technology Laboratories and finally the Stennis Space Center. All subsequent S-IC flight stages were built at the Michoud Assembly Facility near New Orleans, Louisiana and hot fired at MSFC and the Mississippi Test Facility. This chapter will cover the history of each facility's construction and the testing it conducted, rather than covering these events chronologically during the Apollo program.

## SANTA SUSANA FIELD LABORATORY TEST FACILITY

North American Aviation started to construct its first rocket engine test stand in the Simi Hills of Santa Susana, California in 1949 to support the Navaho program. This site was initially called the Propulsion Field Laboratory, but later renamed the Santa Susana Field Laboratory.

Ernie Barrett joined the company in 1946 and was not only among the first test engineers to work on the design of Vertical Test Stand One, known as VTS-1, but was also involved in many tests conducted there. As he told this author in 2007:

The first test stand construction was headed up by Doug Crossland, and included John Finger, myself, and others. The way we positioned the first test stand was John Finger and I got in a Jeep and drove through the brush with 50-gallon drums in the back. We stood where the control center was going to be. I pointed and I said, 'Put a drum there, a drum there, a drum there and a drum there,' and that's how we located the test stand. It was little crude but it worked.

7-1 The Bravo II Test Stand at Rocketdyne's Santa Susana Field Laboratory was used for F-1 turbopump testing. In this photograph dated April 16, 1962, workers complete installation of turbopump R014. (Rocketdyne, Frank Stewart Collection)

The trials of the first Soviet intercontinental-range ballistic missile were known to the U.S. military but were classified. Its existence did not become public knowledge until it was used to launch Sputnik on October 4, 1957. The fact that the Soviets had such a powerful rocket spurred Rocketdyne to expand its facilities at Santa Susana. The test facilities supported the Atlas, Jupiter, Redstone and Thor missile propulsion programs. Six specific test groups were built, named Bowl, Canyon, Alpha, Bravo, Cocoa and Delta, involving 18 large hot fire test stands, five component laboratories having over 60 test positions, and an advanced propulsion test facility. A Horizontal Test Stand (in fact, 30 degrees from horizontal) was also built for Navaho, and later used for other engines. Component Test Laboratory I was built in the mid-1950s to support the testing of rocket engine gas generators, turbopumps, inducers, bearings and seals. Three other Component Test Laboratories were added to support a variety of Gemini, Apollo and unmanned space programs. In the 1960s the Santa Susana Field Laboratory employed as many as 6,000 people, and the facilities often were in use around the clock. The E-1 engine tests were conducted on the Bravo and Delta stands. A variety of facilities tested the thrust chamber, turbopump, fuel and oxidizer valves, high pressure ducts, gas generator, heat exchanger and other components of the F-1 engine. The F-1 precursor known as 'King Kong' was tested on the Bravo stand. Although short duration F-1 thrust chamber tests were mainly conducted on the Bravo stand, long duration full thrust tests of this engine could not be conducted at the Santa Susana Field Laboratory due to the noise and the proximity of housing; this was done at the Edwards Field Laboratory. F-1 component testing continued at the Santa Susana Field Laboratory for the entire engine program.

As with any intensive test program, the men and women who worked at the Santa Susana Field Laboratory often remember the years there in the 1950s and 1960s as the best and most rewarding of their lives. The hours were long, and during periods of aggressive development and acceptance testing it was common to work without days off. Everyone was aware that they were part of something truly historic and momentous. As Clement Cecka (brother of William Cecka) told a Los Angeles Times reporter in 1997, "We were just a new bunch of kids, fresh out of the service, anxious to get a hold on the world. The camaraderie was unbelievable."

## THE EDWARDS FIELD LABORATORY

Given the impracticality of long duration full thrust testing of the F-1 engine at the Santa Susana Field Laboratory, NASA, Rocketdyne and the Air Force decided that the most suitable test site was the Edwards Field Laboratory, sometimes referred to as Edwards Rocket Base, in the high desert of California. Test Stand 1-A had been built by the Los Angeles District of the U.S. Army Corp of Engineers and activated by the Air Force for Atlas missile testing in March 1957. It was initially rated for thrusts of up to 1 million pounds. A portion of the steel superstructure was destroyed when an Atlas exploded undergoing a captive test. The stand was later converted to development and acceptance testing of the F-1 engine. In early 1959, the Army Corps of Engineers was directed by the Advanced Research Project Agency and NASA to

7-2 Fuel pump model 1 installed in Rocketdyne's Component Test Lab No. 5 on November 9, 1962. (Rocketdyne, Frank Stewart Collection)

consult Rocketdyne and provide design support and construction services for the dual-position Test Stand 1-B.

But these facilities alone were insufficient for the aggressive test schedule for the F-1 engine. In addition to development stands, NASA also required test stands for acceptance testing of production F-1 engines. In October 1961 NASA directed the Army Corps of Engineers to draw up preliminary plans and design criteria for three new test stand sites at the Edwards Field Laboratory, including a control center and all related facilities. The test stands were designated 1-C, 1-D and 1-E. The Corps of Engineers selected the Ralph M. Parsons Company of Los Angeles as the principal Architect-Engineer. Rocketdyne's Facilities Engineering group and Test Subdivision provided the design criteria, and input for the on-stand piping, thrust systems, flame deflectors, instrumentation and operations. Larry Skogstand, the company's Chief Facilities Engineer, was Project Manager, Art E. Moore was Liaison Representative, Jim. A. Bowman was Project Engineer, and R. F. Burlew was Project Designer. The design began in January 1962. Initial site surveys began in March, and all phases of design and engineering were completed by June 20, 1962, with the result similar to

7-3 An F-1 heat exchanger undergoing test in Rocketdyne's Component Test Lab No. 3 on November 12, 1962. (Rocketdyne, Frank Stewart Collection).

1-B. The specification included each test stand's superstructure, control building, instrumentation tunnels, electrical support buildings, pre-test buildings, observation bunkers, and both on-stand and off-stand propellant systems. The Corp of Engineers designed all of the excavations and the access road to the acceptance test complex. It also provided a design for the development of water wells and the water supply lines to provide the water needed to cool a test stand during F-1 engine hot fire testing. Construction of the three test stands commenced almost simultaneously. Completion of the first stand and its facilities was scheduled for October 1963, with completion of the others occurring two months and four months later respectively. The total cost for the new facilities was estimated at 30 million dollars at the time.

The 1-C, 1-D and 1-E test stands were spaced approximately 100 yards apart, and were rated for 2.5 million pounds of thrust. They were of an identical design. The reinforced concrete base, made of 10,000 cubic yards of poured concrete, measured 100 feet by 150 feet and stood about 175 feet from the bottom of the concrete base to the top of the moving crane. On top of the steel superstructure were the spherical

7-4 F-1 thrust chamber tests were performed on Test Stand 2-A, also referred to as the horizontal test stand, at the Edwards Field Laboratory. (Rocketdyne, Vince Wheelock Collection)

7-5 F-1 development and production engine acceptance test stands were built at the Edwards Field Laboratory. Two of the stands are shown here on Rocket Ridge. (Rocketdyne, Vince Wheelock Collection)

propellant tanks. The RP-1 fuel tank held 65,000 gallons, and the LOX tank held 90,000 gallons. The LOX tanks had vacuum jackets to minimize evaporation of the LOX held at −297 degrees F. In addition, each stand had tanks for 2,400 cubic feet of gaseous nitrogen. The RP-1 and LOX arrived at the Edwards Field Laboratory by tanker truck and offloaded at a storage tank farm that could hold 420,000 gallons of RP-1 and 670,000 gallons of LOX, and when needed they were pumped to the test stand's propellant tanks; the entire length of the LOX delivery pipeline was vacuum-jacketed. Special equipment on these new stands included heavy load cells to record the thrust of the F-1. Each load cell was accurate to 0.5 percent. Each stand was also fitted with smaller load cells to record the engine's side load thrust. Each stand had a special device to simulated vehicle 'spring rate.' This piece of apparatus lent a more accurate simulation of flight conditions during tests. On Test Stand 1-E there was an environmental chamber surrounding the test position for the F-1. The chamber was capable of chilling the engine to as low as zero degrees F, or of heating it to as much as 130 degrees F. The water for the flame deflector was stored in central vertically mounted cylindrical tanks with capacities of 400,000 and 3 million gallons, and was fed by a pump house to each test stand via a 2 meter diameter pipeline. It could be supplied from the smaller tank at a rate of 22,500 gallons per minute, and from the larger tank at 100,000 gallons per minute. Water used to cool the F-1 engine flame deflector was reclaimed by means of diversion channels and a runoff reservoir. For many of the people who worked and visited this facility, it became known as Rocket Ridge.

## Testing the F-1 at the Edwards Field Laboratory

The Apollo schedule was so tight that for a while the Edwards Field Laboratory was performing development testing and production acceptance testing of the F-1 engine in parallel. Additional hot fire tests of the production engines, and of the S-IC stages in which they were installed, would be done at MSFC and at the Mississippi Test Facility. The development and production acceptance testing undertaken at Edwards was defined as part of the contract negotiation between NASA and Rocketdyne that were related in Chapter 3.

Construction of the first F-1 thrust chamber test facility, designated 2A-2, started in early 1959 under Air Force Facilities Contract AF33 (600)-26940, and it this was activated in January 1960. The first hot fire test of an F-1 thrust chamber using 2A-2 was in December 1960. Thrust chamber tests did not use a turbopump. Rather, they were pressure fed. Propellants were delivered to large, heavy walled ground storage tanks at the stand, and fed to the thrust chamber at high pressure. These tanks were made of stainless steel plate five inches thick. The propellant volume limited thrust chamber tests to 15 seconds. The greatest thrust attained in such thrust chamber tests was measured on April 6, 1961 at 1.64 million pounds of thrust. This record was not broken until the advent of the F-1A development engine discussed in Chapter 9. The successful thrust chamber tests were followed by short duration tests of the full F-1 with nozzle extension in July 1961. The first flight duration test of 2.5 minutes at the engine's full thrust of 1.5 million pounds was conducted on May 26, 1962.

7-6 F-1 development engine No. 007 on Test Stand 1-A on July 9, 1962. Note the configuration of the heat exchanger exhaust manifold relative to later designs. (Rocketdyne, Frank Stewart Collection)

As engines neared final assembly, they were scheduled for testing at the Edwards Field Laboratory, and transported there from Rocketdyne's Canoga Park production facility by truck. Once there, an F-1 was fitted with instrumentation connections and lifted into position on the test stand by using specially designed handling equipment that was capable of moving laterally, rotating the engine through a 90 degree arc and elevating it by almost 2 meters. The F-1's gimbal assembly was bolted to the mating block on the stand, and the hydraulic actuators to gimbal the engine in testing were fitted. Finally, the propellant feed lines were connected and all the instrumentation cable connectors attached. Each engine test was the responsibility of the chief test engineer. Working with him in the control center were the assistant test engineer, who served as the test conductor, an engineer who was responsible for the engine's instrumentation, and numerous development engineers. The lead test stand engineer ran the test stand crew of seven test mechanics and three instrumentation mechanics. The test stand crew would evacuate the test stand area 15 minutes prior to the start of the engine test; every person had to be accounted for prior to proceeding with the countdown. One or two crew members would take up a position in a nearby blast-proof 'pillbox' which had an observation window of plate glass several inches thick. The control center crew monitored the test by closed-circuit television. If the engine developed a combustion instability, this would be detected by a special sensor which would immediately shut down the engine. The close proximity observers had cut-off switches to end the test if they noted any other kind of malfunction. Upon ignition and quickly reaching mainstage operation, the F-1 consumed two tons of LOX and one ton of RP-1 *per second.*

## Test data acquisition and processing

The purpose of F-1 engine hot fire testing was to gather performance data of many different parameters in order to determine rated engine performance and reliability. A vast amount of data was acquired during a test. This was recorded in the control center on as many as 481 channels for a single test (the average was 350). The data was delivered in real time by microwave link to Rocketdyne in Canoga Park, and as necessary to three other North American Aviation data reduction centers in the Los Angeles area. Within an hour the results of the analysis were beamed back to the test site for review by the development and test engineers. The company had the most advanced computers available at that time. The largest single block of data channels was stored in a state-of-the-art digital recording system which sampled the incoming analog signal 15,625 times per second and created digital files for processing. At the same time, the data was recorded on multiple channels at a time resolution as fine as 1/40,000th of a second. This high frequency was necessary to record the myriad of events occurring during the engine's propellant combustion. Other channels were available at medium frequencies on a variety of instruments and ink graph recorders in the control center for real time monitoring. The parameters included the engine's thrust, propellant flow, combustion phenomena, temperatures, pressures, vibrations,

7-7 Rocketdyne technicians work on an F-1 engine in one of the dual-position test stands at the Edwards Field Laboratory. Engines were tested with and without the nozzle extension—in this case without. (Rocketdyne, NASA/MSFC)

side loads and valve operations. On completion of its hot fire tests, each engine was returned to Canoga Park for any component changes and other corrective actions. In some cases an engine was disassembled and inspected. Once the modifications had been completed, the engine would be sent back to the Edwards Field Laboratory for further testing.

As Richard Brown of MSFC's F-1 engine program office told this author:

> The philosophy used to achieve the needed reliability was extensive engine systems testing under varying conditions, such as 'bomb' tests for combustion

stability, off-nominal mixture ratio tests, turbine over-speed tests, and tests at the upper and lower operating temperatures. In addition to the development testing, there was a formal flight qualification program and acceptance testing of production flight engines for shipment. The most basic driving parameter in terms of saying the engine was ready for launch, was the number of hot fire tests and test seconds on the engine without a premature cutoff or hardware failure. At one point, Dr. von Braun was concerned the hot fire testing rate was too low to have confidence in the reliability of the flight engines. He went out to Rocketdyne to talk to Sam Hoffman, the president of the company, about his concerns. He told Mr. Hoffman, in so many words, 'You've got to speed up the testing on the F-1 or we are not going to be ready to fly on time.' This senior management to senior management 'chat' resulted in a thirty to forty percent increase in the testing rate. After about a year of this, we decided to incentivize—that is, we would pay an extra fee if they could reach a certain testing rate [and] low and behold the testing rate went up by perhaps another forty or fifty percent!

In those days, we didn't have the ability do the analysis and simulations that we can do today with computers. We had to 'bang out' the tests—what some have called attribute testing—to verify the flight readiness of the engine. This meant we hot fired the engine repeatedly, each time measuring its attributes such as chamber pressure, pump speed and performance, combustion stability, temperatures, flow rates and things like that. At the height of development program, in the 1965–1966 timeframe, we were averaging some sixty to sixty-five tests per month. We finished the formal flight qualification test program in the fall of 1966. In the year prior to completing the qualification program, I think we ran about 800 tests. Rocketdyne did an outstanding job in achieving these testing rates.

I remember once we had a tough negotiation over the testing rate needed for them to earn their incentive fee. We went back and forth with Rocketdyne several times, and finally agreed the number of tests. Their Assistant Program Manager, Dom Sanchini, remarked at the end of the negotiation, 'You realize, don't you, that the LOX for all these tests has to come from Los Angeles and that we're going to have LOX trucks bumper to bumper from Los Angeles to Rocket Ridge?' We didn't take his remark seriously at the time. One day not long after that, I was sitting in my office when one of our resident engineers out at Edwards, J.M. Smith, called. He chuckled, and said something to the effect of, 'I was up on Rocket Ridge. I looked to the west and there were LOX tanker trucks coming—a convoy as far as I could see. That's something to see. It looks like Dom was right. We must have all the LOX in southern California coming up here.'

I was told on another occasion one of the LOX truck drivers pulled up and unloaded his LOX. About that time there was an engine firing. After the firing he asked one of the test people 'About how many tests will you get out of this load?' The test engineer said, 'Well, it takes eleven or twelve truck loads like yours for one engine firing.' It was said that the driver looked amazed and perhaps a little bewildered.

7-8 Testing an F-1 without its nozzle extension on Test Stand 1-A at the Edwards Field Laboratory. (Rocketdyne, Edwards AFB Main Base History Office)

## MSFC TESTING

The Test Laboratory at MSFC was a carry over from its ABMA days. In addition to the facilities for testing engines for the Redstone and Jupiter missiles, there were test stands for the Saturn I powered by a cluster of H-1 engines. The contract issued by the Advanced Research Project Agency for the single stage Saturn I was followed up by a supplemental to develop a multiple stage Saturn I. On December 11, 1958 ARPA order 47-59 authorized the Army Ordnance Missile Command in Huntsville to design, construct and modify the Captive Test Tower that had been built in 1951 and used in testing the Jupiter. On January 14, 1959 ABMA initiated construction of this Saturn I Static Test Stand. When completed later in the year, it towered 177 feet. It was a dual-position test stand; the west side to test the S-I stage and the east side to test the S-IB that was to follow.

### F-1 single engine test stand construction

To permit early testing of the F-1 at MSFC while the design and construction of a dedicated F-1 single engine hot fire static test stand was underway, in February 1963 work began to adapt the west side of the Saturn I test stand to accept the F-1 engine. The first hot fire test of an F-1 in this modified part of the dual-position test stand was on December 3, 1963, and lasted for 1.25 seconds. After a thorough inspection of the engine, a second firing was made for a duration of 10 seconds. The dedicated F-1 Engine Test Stand was completed in September 1964. It was 4,600 square feet in area at the base, stood 239 feet tall, and its foundations were anchored to bedrock some 40 feet beneath ground level. It was rated for 1.5 million-plus pounds of thrust, but with modifications would be able to accommodate engines delivering thrusts of 3.0 million pounds. All control and instrumentation requirements were provided by the West Area Blockhouse. The deflector was cooled by water flowing at a rate of 135,000 gallons per minute. The overhead derrick had a lifting capacity of 100 tons. MSFC also had F-1 engine component test stands, including one for the turbopump.

Dr. William Mrazak directed the Propulsion and Vehicle Engineering Laboratory, with Herman Weidener as his deputy. Hans Paul was manager of the F-1 Propulsion Division of the laboratory. Ron Bledsoe, who was the F-1 Project Engineer in this division, was responsible for coordinating the engineering inputs to the F-1 testing at MSFC. He vividly recalled his dealings with personnel in the Test Laboratory:

> The Test Lab people were tough as nails. Karl Heinberg was director during the Apollo days. He wouldn't even let Dr. Arthur Rudolph, the manager of the Saturn V Program Office, out there to view the tests. Heinberg told Rudolph, 'If you don't have a reason to be out here, I don't want you there.' Rudolph said, 'All I want to do is watch the test,' and Heinberg responded, 'I'll tell you what happens with the test!'

The F-1 Engine Program Office also included reliability engineers, and as Richard Brown recalled:

> We had three reliability engineers assigned to the F-1 Engine Program Office. They monitored the test results, and made reliability calculations. Their work also helped us maintain visibility of overall progress toward a reliable, human-rated engine. They were Woody Bombera, Hal Steinberg, and Sid Lichman. [...] They looked at all the F-1 engine component and engine system test data, and main-

7-9 Insulation protected the F-1 engine both at ignition and against base heating in the upper atmospheric phase of the S-IC's powered flight. (Rocketdyne, Sonny Morea Collection)

7-10 The first production F-1 engine, F-1001, was delivered by truck to Test Stand 1-B-2 at Edwards Rocket Test Site on September 28, 1963 for acceptance testing. (Rocketdyne, Harold C. Hall Collection)

tained running charts on successes, failures and premature cutoffs. They received failure and premature cutoff reports from the contractor in both component and engine system testing, and cataloged the reasons and failure modes. They used this information to make predictions of engine maturation. What we were interested in, was isolating any potential problems and getting premature engine cutoff versus [number of] tests and test-seconds [in order] to get the premature cutoff rate down to as near zero as possible.

## The S-IC stage static test stand

MSFC needed the capability to test the S-IC first stage of the Saturn V all the way from ignition of the five F-1 engines to shutdown after the full duration of mainstage operation. In addition, tests had to be made to validate the performance of the other systems on the S-IC, and make any necessary design changes. The Test Laboratory issued the design specifications for the S-IC Static Test Stand. Brown Engineering in Huntsville provided engineering and design support to MSFC. A civil engineering construction firm in San Francisco, California won the contract to build the massive four column concrete base, with foundations going down 40 feet to the bedrock. The contract for fabrication of the superstructure was issued on August 31, 1962. Overall construction was supervised by the Mobile District of the Army Corps of Engineers. The concrete portion of the test stand was finished by July 1963, at which time work began on the steel superstructure. The test stand was capable of withstanding a thrust of 12 million pounds (almost 54 MN). With the crane set atop the superstructure, the stand rose 122 meters. Large storage tanks on the ground, and smaller tanks on the stand, fed LOX and RP-1 to the stage. The steel flame deflector was cooled by water pumped from the Tennessee River and held in local tanks. The stand was completed in 1964 at a cost of $30 million, and it would be used to test the S-IC-T development non-flight stage, as well as several flight stages.

Karl Heimburg was Director of the Test Laboratory. In 1960, he assigned Harry M. Johnstone, who was a veteran of missile and rocket tests since the early ABMA days, as Test Director for the Saturn I and, beginning in 1964, for the S-IC stage of the Saturn V. Joe Lundy, John Funkhouser, Ron Tepool, Jan Monk, Al Dawly, Tom Shaner, Carl Fuller, John Odom and Fred Cunningham were among the engineers and technicians engaged in F-1 engine systems installation, checkout and operations. Corry Coleman, Henry Ferrieo and Bob Christian were among the instrumentation engineers. Rocketdyne had a resident engineering office at MSFC's Test Laboratory, as well as in the manufacturing and quality laboratories, to provide a technical link between MSFC F-1 engine operations and Rocketdyne in Canoga Park. As noted by Vince Wheelock, the company's Resident Manager at MSFC, this function assisted NASA in remaining current and implementing changes resulting from testing at the Santa Susana Field Laboratory and the Edwards Field Laboratory. The Rocketdyne resident supply function at MSFC was active in providing the provisioning, issuing and warehousing of F-1 hardware, and it could even order spare parts, assemble, and test two F-1 spare engines.

7-11 A U.S. Air Force C-133 Cargomaster was used to deliver F-1 engines to the Marshall Space Flight Center. (Rocketdyne, Vince Wheelock Collection)

## Testing the S-IC-T stage

MSFC and Boeing built the S-IC-T static test stage for the purpose of validating the compatibility of the stage with the F-1 engines and the interfaces between the stage and the test stand. S-IC-T arrived at the test stand on March 1, 1965, without its F-1 engines and suction ducts. The initial schedule called for the first static test firing on June 15, but Johnstone and his team installed the engines, propellant ducts, electrical and hydraulic connections, and completed the systems checks nine weeks in advance of schedule, enabling the first test to be conducted on April 9. This single F-1 engine test was not without its problems, but this was expected. However, there was a hair-raising discovery. As Harry Johnstone wrote years later:

> Cutoff was given by an observer inadvertently. We recycled the count, and fired two and a half hours later. Cutoff this time was given by main fuel valve No. 1 safety circuit, which indicated the valve closed, but actually it was open. After the test, we found that four of the eight bolts [fixing] the engine gimbal block to the thrust structure had broken. Bolts of the wrong material had been used. One more broken bolt, and we possibly could have had a catastrophic situation if the F-1 engine had fallen from the vehicle. A fire and an explosion would have followed. We static fired all five F-1 engines on the 7.5 million pound thrust vehicle, the largest in the world, on April 16, 1965; two months ahead of schedule.

This first five-engine test firing of the S-IC-T stage lasted for 6.5 seconds. During this time, over 500 measurements of the engines and the S-IC-T were recorded. The sound of five F-1 engines operating at mainstage was the loudest noise ever heard in Alabama; being heard tens of kilometers away. It was a milestone in the F-1 engine program, the Saturn V program, and the Apollo program. Those who witnessed this test were awed by the power of five F-1 engines firing simultaneously.

As Dominic Sanchini, Rocketdyne's F-1 engine assistant program manager, said, "It was the most spectacular sight I've ever seen. After working on the F-1 program for six and a half years, it was extremely gratifying to see the first F-1 cluster test."

Also observing were Vince Wheelock and William Rance of Rocketdyne's Field Engineering office, and James Owlsley of its Contracts and Proposals office. It was a moment in American rocket history as significant as, perhaps, the day of the Trinity atomic bomb test. After fifteen successful static firings at Huntsville, S-IC-T was transported to the Mississippi Test Facility in early 1966 in order to validate the new test stand there. As Harry Johnstone wrote of this occasion:

> I gave thanks and praise to the greatest test team in the world in meeting this great milestone. Dr. von Braun was elated. I had a dream team behind me, and there was no way we wouldn't succeed. My team gave their heart and soul to the program, working long, stressful and sometimes dangerous hours. I might add here that in motivating my engineers and technicians, we would stack our hands on a table, one on top of another, drawing the conclusion that we were all one together with the single purpose to meet our goal. There was an air of confidence about the whole team.

Three S-IC flight stages were tested at MSFC. The first two were built at MSFC by a joint MSFC and Boeing team. S-IC-1 completed post-manufacturing checkout in January 1966. It was transported to the Static Test Stand, lifted into position and secured on January 24. The first test firing was on February 17, and the programmed duration of just under 41 seconds met all test objectives. Another test was conducted on February 25, but a red-line observer reacted to an incorrect reading by ending the test 83.2 seconds into the planned duration of 128 seconds. The incorrect reading proved to be due to a faulty transducer. After analysis of the results, it was decided that a full duration test would not be needed. S-IC-1 was refurbished, subjected to a number of inspections and checks by MSFC's laboratories, loaded on the sea-going barge *Poseidon* and shipped to KSC. It became the first stage of AS-501, which was launched unmanned as Apollo 4 on November 9, 1967.

S-IC-2 was installed in the Static Test Stand at MSFC in May 1966. The first test firing on June 7, 1966 lasted for 126.3 seconds. However, one of the F-1 engines on this stage, F-4017, showed thrust anomalies. After being removed it was inspected, refurbished and subjected to three further individual tests during which it functioned satisfactorily. It was reinstalled on S-IC-2 in August. Refurbishment, inspection and checkout of the stage by several of MSFC's laboratories took several months. It was then loaded aboard the barge *Poseidon*, shipped to KSC in March 1967, and became the first stage of AS-502, which was launched unmanned as Apollo 6.

7-12 The dual-position Saturn I test stand at the Marshall Space Flight Center was modified in 1963 to permit testing of the F-1 engine prior to the completion of the dedicated F-1 test stand. (NASA/MSFC)

The first S-IC to be manufactured by Boeing at the Michoud Assembly Facility in Louisiana was S-IC-3, and it was the final stage tested at MSFC. It was shipped by the barge *Poseidon* from Michoud to MSFC and placed on the Static Test Stand on October 3, 1966. The first test on November 15, 1966 lasted for 121.7 seconds. The stage was removed from the stand on November 21, and returned to Michoud. After refurbishment, engineering change orders and systems checks, NASA accepted it for shipment to KSC. On arrival there on December 27, 1967, it was placed in storage. It became the first stage of AS-503, which was to be the third unmanned test flight of the Saturn V, but when this was cancelled the vehicle was used for Apollo 8, the first manned mission, which flew out to orbit around the Moon.

No further tests of S-IC flight stages were made at MSFC. All subsequent stages were sent to the Mississippi Test Facility.

## MISSISSIPPI TEST FACILITY

During the studies in 1961 regarding launch vehicles and the mode by which Apollo was to fly the Moon, the launch vehicle of choice for the direct ascent option was the Nova, powered by eight F-1 engines. Because the noise from testing such a first stage was unlikely to be tolerated by the citizens of Huntsville, in June 1961 NASA and the Department of Defense undertook joint studies of sites at which the stages of the lunar launcher could be tested. At the same time, NASA had to decide where this vehicle would be manufactured and where it would be launched. A key requirement was that the facilities for first stage production, hot fire testing and launch should be relatively close to each other and readily accessible from the sea. In August, NASA announced that the launch site would be at Cape Canaveral, Florida. In September, it selected the dormant defense plant near Michoud, on the coast of Louisiana, as the manufacturing site for the first stage. For the hot fire acceptance test facility, a test site selection committee made up of personnel from NASA Headquarters and MSFC met on August 7, 1961. Owing to the noise that testing would create, the site had to be far from populated areas, which meant an inland location. But it also had to have access to a river to enable barges to ferry tested stages back to Michoud for revision or to Florida. In addition, the climate had to be conducive to year-round testing. The list of potential sites was soon cut to six places in Louisiana, Texas, Mississippi and Florida. After on-site inspections, the committee employed a points system to select a 13,800 acre tract in Hancock County, Mississippi, alongside the Pearl River, as the primary test site, with approximately 128,000 additional acres for a buffer zone. On August 26 this recommendation was sent to James Webb, the NASA Administrator. On October 25 the agency announced the selection. The choice came as a surprise to many. Critics charged that the decision had been influenced by Mississippi Senator John C. Stennis. Dr. von Braun wrote to *Aviation Week* editor George Alexander and said that the long term reasoning behind selecting the Pearl River site was to have a facility that could test not only the rockets that were on the drawing boards but also the larger vehicles that might be needed in the next 25 to 50 years, which meant the facility would have to be designed for thrust levels of 20 million pounds.

### Building the Mississippi Test Operations test stands

NASA again selected the U.S. Army Corps of Engineers to execute the massive civil engineering project for what was initially called Mississippi Test Operations, but in 1965 became the Mississippi Test Facility. The first step was to obtain the privately

7-13 The dedicated F-1 engine test stand in the west test area of the Marshall Space Flight Center was used throughout the F-1 engine program. (NASA/MSFC)

held land, both individual and corporate. The process of informing the land holders of the intention to compensate them for their land was necessary but understandably difficult. Some families had held land for generations, and were reluctant to leave. The reasons for people having to move were simple. The noise in the test area would be as much as 125 decibels, which would be lethal, and even the 110 decibels in the buffer zone would be hazardous to health. In fact, the low frequency pressure waves could even destroy buildings. For that reason, many of the homes were relocated to other parts of the county. In 1963 NASA hired Sverdrup, Parcel & Associates of St. Louis, Missouri for master planning and architectural design. Of the NASA centers, MSFC was the most actively involved, helping with its management organization. A network of access waterways for the barges was the first priority, and in excess of 20 kilometers of the Pearl River were dredged to the required depth and width. More than 10 kilometers of canals were also dug within the Mississippi Test Facility, and a lock system built to allow access to the Pearl River. The Army Corps of Engineers set up their preliminary offices in Cypress House, formerly a lodge, located near the Rouchon House where NASA personnel and other individuals would initially work.

When NASA selected the Saturn V to launch Apollo, the A-1 and A-2 test stands at the Mississippi Test Facility were built for the S-II second stage, and the dual-position B-1/B-2 test stand was built for the S-IC first stage. The latter was secured using over 1,500 steel pilings each 30 meters long, and it stood 124 meters tall.

## S-IC stage testing at the Mississippi Test Facility

Rocketdyne provided in residence operations engineering, technicians and quality functions at the Michoud Assembly Facility and the Mississippi Test Facility able to undertake 'hands on' work during F-1 engine and S-IC stage processing. This team was headed by Rosco Nicholson, who joined Rocketdyne in the early days at MSFC, moved to manage the company's operations at Michoud, and later served in this role at the Mississippi Test Facility.

The S-IC-T stage reached the Mississippi Test Facility on October 23, 1966 and was placed in storage while the B-2 stand was completed. It was raised into position and secured to this stand on December 17, and then the stage, its engines and the test stand itself underwent several months of checkout. The first test firing on March 3 1967 lasted 15.2 seconds. The second test on March 17 was for just over 60 seconds, and it validated the B-2 test stand. The stage was removed on 24 March and shipped by the barge *Poseidon* to MSFC in April, where it was returned to the S-IC Static Test Stand for a final test designed in part to train KSC launch personnel.

The first flight S-IC stage to be acceptance hot fire tested at the Mississippi Test Facility was S-IC-4. It was shipped by barge from the Michoud Assembly Facility in April 1967, and subjected to several weeks of checkout on the B-2 stand. After fuel loading on May 15 and LOX loading on May 16, which was the day of the first test, the five F-1 engines roared to life and ran for the full duration of 125 seconds. There were a number of anomalies during this test, but they were not deemed critical. The stage underwent thorough post-firing checks, was removed from the stand on June 7, and loaded onto the barge *Pearl River* for shipment to Michoud for refurbishment.

7-14 Rocketdyne, Boeing and Marshall Space Flight Center personnel complete installation of the fourth outboard F-1 engine on an S-IC stage inside the S-IC Static Test Stand at MSFC in March 1965 in preparation for a full stage test the following month. (NASA/MSFC)

The checkout at Michoud established that a number of components were in need of rework or replacement. After another round of tests and inspections, it was placed in storage. On being removed from storage in November 1967, Boeing made a number of modifications. It was again checked and systems tested in the Michoud Assembly Facility's Stage Test Building. Although NASA accepted the stage in March 1968, it had to undergo further tests later in the year. It finally left Michoud aboard the barge *Orion* on September 24, 1968. S-IC and S-II stages shipped by barge made their way into the Gulf of Mexico, around the tip of Florida south of Key West, up the eastern coast to the port at Cape Canaveral, then up the Banana River to the KSC turnaround basin dock for off-loading onto wheeled transport. S-IC-4 arrived on September 30, and became the first stage of SA-504 which was launched as Apollo 9.

The S-IC-5 stage was secured to the B-2 test stand on June 29, 1967. It underwent nearly two months of preliminary firing tests and checkouts, and on August 25 was fired for the full duration of just over 125 seconds. Post-test inspection revealed an engine thermocouple lodged in the throat of one of the F-1 engines, and a broken turbine inlet transducer. It was returned to Michoud in September for refurbishment and then planned modification in between periods of storage during the rest of 1967 and much of 1968. It was loaded onto the barge *Orion* in November 1968 and taken to KSC. Upon arrival there, it underwent another round of inspections and storage preparations while awaiting stacking as SA-505, which was launched as Apollo 10.

On March 1, 1968, S-IC-6 arrived at the Mississippi Test Facility on the barge *Pearl River*. It was installed in the B-2 stand on March 4 and subjected to months of checkout and tests. Finally, on August 13, it was fired for the planned 125 seconds. All five F-1 engines performed within their performance parameters. This test was particularly significant due to the inclusion of the 'pogo' suppression system, added after Apollo 6 suffered from this longitudinal oscillation when launched on April 4. Following the post-firing checks, the stage was returned to Michoud on August 28 for refurbishment, modification and checkout. On 16 February 1969 it was loaded onto the barge *Orion* for shipment to KSC, where it arrived four days later. S-IC-6 is perhaps the most historically significant of the Saturn V first stages because it was stacked as SA-506 and launched as Apollo 11, the mission that achieved Kennedy's challenge of landing men on the Moon.

The S-IC stage was proving to be a reliable booster, but this was primarily due to the careful manufacturing, meticulous inspection, methodical testing and thorough documentation.

S-IC-7 was shipped from Michoud to the Mississippi Test Facility on September 12, 1968 and installed on the B-2 stand. During the extended countdown, a leak was noted in the fuel tank. The tank was drained and the leak fixed. The rescheduled test on October 30 ran for the planned duration of just over 125 seconds. The stage was removed from the stand on November 8, and shipped back to Michoud the next day. On the completion of refurbishment and checkout in late January 1969, it was placed into temporary storage, then loaded on the barge *Orion* and set off for KSC on April 29. It was stacked as SA-507 and launched as Apollo 12. A curious thing happened with this mission. After a decade of monumental effort and mounting expectation, the achievement by Apollo 11 of the first manned lunar landing on July 20, 1969 not only marked America's finest hour, it also produced a sense of anticlimax that made public interest wane. However, those building and testing the Saturn V continued to prepare their vehicles with enthusiasm and professionalism.

Assembly and checkout of S-IC-8 at Michoud was finished by the end of October 1968. It was shipped to the Mississippi Test Facility on November 13 and placed on the B-2 stand. The test firing took place on December 18, and ran to the full duration of 125 seconds. It was removed from the stand on January 3, 1969, and shipped to Michoud the same day. As before, the scheduled refurbishment, inspection, tests and checkout involved months of work. The stage was shipped on June 11 and arrived at KSC on June 16. It was stacked as SA-508 and launched as Apollo 13.

The S-IC-9 stage which arrived at the B-2 stand at the Mississippi Test Facility on January 10, 1969 was the result of two years of manufacturing, testing, modification and checkout—as indeed was the case for almost every such stage. On February 19 it fired for the full 125 seconds. After post-firing checks, it was placed on a barge on March 5 for return to Michoud. It underwent refurbishment and further static

7-15 The first test firing of an S-IC stage with all five F-1 engines running was on April 16, 1965. The S-IC Static Test Stand at the Marshall Space Flight Center permitted development and flight stage qualification testing prior to shipment to the Kennedy Space Center. (NASA/MSFC)

non-hot fire testing, and was then placed in storage. In December it was prepared for shipment to KSC, where it arrived on January 12, 1970. Just over a year later, it was stacked as SA-509 and launched as Apollo 14.

At any one time, there were several S-IC stages at different phases of assembly at Michoud. Such was the case with the next three in October 1968. S-IC-10 completed post-manufacturing checkout in the first week of March 1969 and was shipped to the Mississippi Test Facility on March 11. It was installed on the B-2 stand the next day. Although the test on April 16 ran for the full 125 seconds, one of the F-1 engines gave less thrust than expected. The stage was shipped to Michoud on May 2. After the suspect F-1 engine had been inspected and analyzed, it was replaced by a flight qualified spare. Following complete stage refurbishment and checkout, S-IC-10 was put in storage. It was shipped on July 1, 1970, arriving at KSC on July 6. This stage had the recalibrated F-1 engines required for the increased payload to permit a lunar module to land with the Lunar Roving Vehicle and stay on the Moon for three days. This recalibration was achieved by a retrofit kit which resulted in 1,522,000 pounds of thrust at sea level, compared to the original 1.5 million pounds of thrust. S-IC-10 was stacked as SA-510 and launched as Apollo 15, which was the first of the J-class missions.

S-IC-11 provided a surprise for NASA, Boeing and Rocketdyne. It was completed at Michoud in May 1969 and shipped to the Mississippi Test Facility, where it was installed on the B-2 stand. Almost immediately, it displayed a number of problems, including a fuel system leak. Finally, the firing test was scheduled for June 29. The ignition sequence commenced and the F-1 engines started smoothly. But at just over 96 seconds into the test a fire started owing to a fuel leak at the supply line to the thrust vector control gimbal. The stage was shut down, and the fire extinguished by the water of the fire suppression system. On removing the stage from the stand, it was found that two of the engines had been so badly damaged by the fire that they would have to be scrapped. The remaining engines required a thorough inspection and refurbishment. At Michoud, all of the engines were removed from S-IC-11 and replaced with new ones. However, the complete stage repair and refurbishment took a year. It was returned to the Mississippi Test Facility and installed on the stand on May 7, 1970. The test on June 25 was halted 70 seconds into the planned duration of 125 seconds owing to a high LOX ullage reading. Despite this, because all the other test milestones had been achieved it was decided not to attempt a full duration firing. The cause of the high LOX ullage reading was identified, and addressed in the post-firing checkout and refurbishment at Michoud. The stage was placed in storage, then shipped to KSC, where it arrived on September 17, 1971. It was stacked as SA-511 and launched as Apollo 16. Despite its test problems, S-IC-11 performed flawlessly in flight.

After completing manufacture at the Michoud Assembly Facility in June 1969 the S-IC-12 stage was shipped to the Mississippi Test Facility in mid-July and installed on the B-2 stand on July 18. The test firing on November 3 was flawless. It was then shipped to Michoud, arriving on November 19. After refurbishment, it was placed in storage until March 1971, when it went through additional systems tests and finally prepared for shipment. The barge left Michoud on May 6, 1972 and

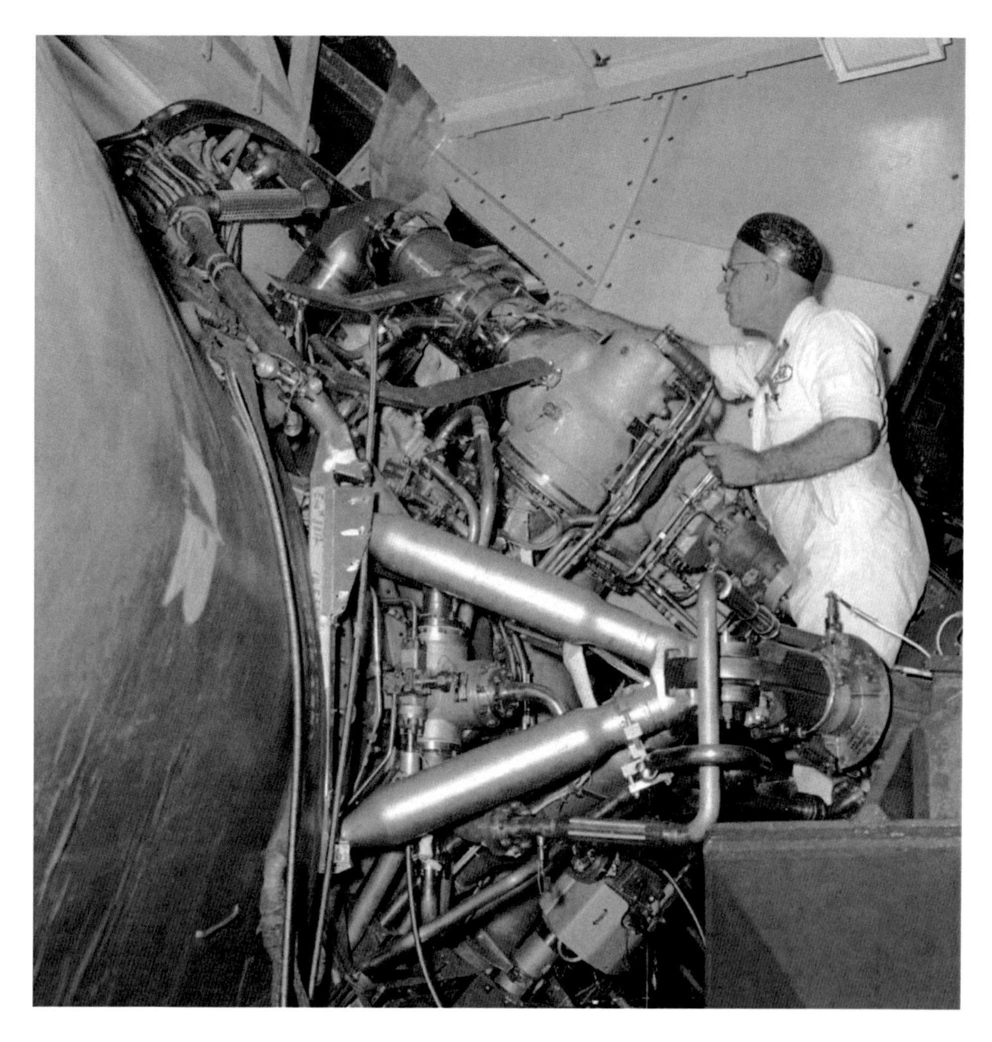

16. Prior to and following test firing of the S-IC stage at the Marshall Space Flight Center, the stage and its F-1 engines were thoroughly inspected by the Quality and Reliability Assurance Laboratory. (NASA/MSFC)

arrived at KSC five days later. S-IC-12 was stacked as SA-512 and launched as Apollo 17—the final manned mission to the Moon.

S-IC-13 was built in the expectation that it would launch an Apollo lunar mission, but budget cuts precluded further missions. However, as a result of a decision made in July 1969 a Saturn V was needed to place America's first space station into orbit, and therefore S-IC-13 was assigned to the Skylab program. The post-manufacturing checkout was completed in October 1969. The stage was shipped to the Mississippi Test Facility on November 20, and installed on the B-2 stand four days later. It went through its preliminary battery of tests and propellant loading for

7-17. An S-IC stage is hoisted by crane for installation on Test Stand B-2 at the Mississippi Test Facility in March 1967. Both the first and second stages of the Saturn V were tested at this facility. (NASA/Stennis Space Center Collection)

the static test firing on February 6, 1970, which ran for the planned 125 seconds. The stage was returned to Michoud for refurbishment and testing. Because the Apollo budget cuts meant that the mission rate had to be reduced, the launch of Skylab was slipped from 1972 to 1973. The barge *Orion* delivered S-IC-13 to KSC in July 1972, and it was launched ten months later.

S-IC-14 was assembled at Michoud in 1969 and completed its post-manufacturing checks in February 1970. It was shipped to the Mississippi Test Facility on March 5, and installed on the B-2 stand. The test on April 16 ran to the planned duration. The stage was removed from the stand on May 1, and returned to Michoud for checkout and refurbishment, then returned to the Mississippi Test Facility for further non-hot fire tests and placed in storage there in August 1970. It was removed from storage in January 1971 and returned to Michoud to complete its systems tests and checkout, these being completed in May 1971. As it was not required at KSC, it was placed in storage. It was eventually taken to the Johnson Space Center in Houston and put on display, mated with upper stages and an Apollo spacecraft. After more than a quarter of a century of deterioration due to exposure to the elements, in 2005 the Saturn V was restored and placed in a building erected specifically to house it.

S-IC-15 underwent its post-manufacturing verification the end of July 1970, was shipped to the Mississippi Test Facility the following month and installed on the B-2 stand on August 17. The test on September 30 ran for the planned 125 seconds. The stage was returned to Michoud and the checkout and refurbishment was completed in March 1971. It was placed in storage as the backup for Skylab, and possibly later to launch Skylab II, if this were funded. It subsequently became the property of the Smithsonian Museum, but remained at Michoud on its original transporter as part of an outdoor display, where it is today.

There is an aspect to all these S-IC stages that bears reflection. After successfully propelling the upper stages of the Saturn V, in each case the stage separated and fell into the Atlantic, experiencing structural damage during the reentry and on impact, with the mangled stage and its F-1 engines sinking to the floor of the ocean, the final resting place of these magnificent machines.

# 8

# KSC and Apollo Saturn

NASA formally announced the creation of Project Apollo on July 28 and 29, 1960 at a Program Plans Conference. Aerospace executives converged on Washington, D.C. for one of the most momentous industrial briefings by any government agency. The program to develop a spacecraft capable of flying three astronauts to the Moon was detailed, the requisite launch vehicles were discussed, the status of current programs was explained, and long range plans were outlined. But the schedule and the budget were ill defined. Then on May 25, 1961 President John F. Kennedy stood before a joint session of Congress and boldly made the daunting task of landing a man on the Moon before the decade was out a national priority. He knew that this task would be technically challenging, but the engine to power the first stage of the launch vehicle, the Rocketdyne F-1, was already under development.

The Apollo program would affect the lives of central Floridians in particular. The residents of Cape Canaveral, Titusville, Melbourne, Palm Bay, Merritt Island, Cocoa and Cocoa Beach learned of Kennedy's speech in the local newspapers the next day. Collectively, these small towns comprised what became known as the Space Coast. Having witnessed launches of Jupiter, Atlas, Titan, Thor and other missiles from the Cape, these communities already knew they were part of a dawning space age, but it was the launch of astronaut Alan B. Shepard on a suborbital flight earlier that month that had really grabbed their attention. The decade of the 1960s would transform the Space Coast unimaginably. Saturn Vs would be readied in a building rising 525 feet, vast crawlers would carry them to concrete pads on the sandy shore, and they would lift off to the thunderous roar of F-1 engines. The thousands of men and women who flooded into the area to work on Apollo gave this place a name: Moonport, U.S.A.

### Launch complex facilities configuration

In the late 1950s and early 1960s, several studies were made of launch facilities for large rockets. One study involved an offshore platform in the style of the oil drilling

8-1 Dr. Kurt H. Debus was Director of Launch Operations for the Mercury, Gemini and Apollo programs, and played a major role in designing the facilities for the Saturn V launch vehicle. (NASA/KSC)

rigs in the Gulf of Mexico nick-named the Texas Towers. This would have had the advantage of not subjecting populated areas to the intense sound pressure levels, but it had many disadvantages and was not adopted.

The means of preparing a large launch vehicle was also the subject of much study. The Saturn I was assembled and tested on the launch pad, but the pace of launches predicted for Apollo was such that it seemed better to undertake all the preliminary work inside a special structure. Dr. Kurt H. Debus, head of the Launch Operations Directorate, was the main architect of the mobile launcher concept, in which a large rocket would be prepared indoors on a transportable launch platform. In February 1961 he met with George von Tiesenhous and Theodor Poppet of his Future Launch Systems Study Office to discuss the infrastructural requirements of the large Apollo launch vehicle. The Future Launch Systems Office presented its findings in April. Included was the transfer of the lower stages by barge from the assembly and testing facilities to the Cape, assembly of the stages inside a large fixed structure, then the transport of the vehicle to the pad in a vertical configuration. Offshore assembly and launch was also studied. After considering the analyses with Dr. von Braun, Debus met the Associate Administrator, Robert C. Seamans, at the end of April. Impressed with the thoroughness of this preliminary work, Seamans authorized a detailed study and cost analysis of the mobile launcher concept—although it would be necessary to conduct parallel studies of launch facilities for a large Saturn and the even larger Nova until the method by which Apollo would fly to the Moon was decided. It was directed by Dr. Debus and Maj. Gen. Leighton Davis. The report, delivered on July 31, was entitled *Joint Report on Facilities and Resources Required at Launch Site to*

8-2 Dr. George Mueller made a presentation to President John F. Kennedy with NASA upper management in attendance in November 1963. The location was the blockhouse at Launch Complex 37, which was configured for the Saturn I launch vehicle. (NASA/KSC)

*Support NASA Manned Lunar Landing Program.* This studied the requirements of: a Saturn C-3 in which all the stages used liquid propellants; a Saturn that had a solid propellant booster and liquid upper stages; an all-liquid Nova with eight F-1 engines in the first stage; and a Nova that combined a solid booster with liquid upper stages. Eight possible launch sites were evaluated. There were four key elements to this new assembly and launch concept:

1. Vertical assembly of the launch vehicle inside a massive building on a mobile launch platform with an integrated service tower.
2. The loaded mobile launch platform would be transported to the pad, where the vehicle would undergo final checkout and propellant loading.
3. Launch operations would be controlled from a remote center within sight of the pad.
4. Checkout of the vehicle and countdown to launch would be automated.

In the summer of 1961 Debus created the Heavy Space Vehicles Systems Office, directed by Rocco Petrone, to define the requirements and estimate the budgets for a Vehicle Assembly Building (VAB) and the associated launch complex. Petrone was assisted by J.P. Claybourne and William Clearman. They monitored the debate over the mission mode and the refinement of the Saturn and Nova concepts, and adjusted their studies accordingly. Meanwhile, in August NASA announced that the facilities would be built at Cape Canaveral. By the end of 1961 the decision was made to add the fifth F-1 engine to the Saturn C-4, as the C-5, and in July 1962 it was announced that Apollo would use lunar orbit rendezvous. With the issue resolved, it was decided to rename America's Moon rocket the Saturn V.

In addition to the Debus–Davis study, the Launch Operations Directorate studied various means of transporting the platform carrying the launch vehicle, including by barge, rail and some form of crawler. A barge would be by far the most expensive approach, with the cost of designing and excavating the basin itself estimated at $20 million dollars at the time. The rail method posed technical difficulties in controlling the engine when following curves, and the cost of a rail bed would be twice that of a road. A crawler would be based on the technology of large earth-moving equipment. In June 1962 the Army Corps of Engineers issued a report saying that Merritt Island was the best part of the Cape on which to build a crawlerway. Later in the month, NASA chose the crawler-transporter approach for the Saturn V, thereby marking another milestone in America's goal of reaching the Moon within the decade.

## LAUNCH COMPLEX 39

With the decision made for lunar orbit rendezvous and assembly of the Saturn V on a mobile transporter in an enclosed VAB, bids were issued for all these critical elements over a period of months during 1962 and 1963. This Apollo infrastructure was identified as Launch Complex 39. On December 3, 1962, the U.S. Army Corps of Engineers, NASA's chosen civil engineering program management organization, awarded design study contracts to four New York architectural firms for the VAB for a total of $3.3 million, and contracts were later issued to the Morrison–Knudson Company, the Perini Corporation and Paul Herdeman Incorporated to construct it. In terms of volume, the cube-shaped structure would become the largest building in the world. Site clearance and ground preparation had started in 1962. The next year saw the steel piles driven in and the concrete foundation slab poured. Construction of the structural steel began in January 1964, and progressed with amazingly rapidity. The VAB was "topped out" (structurally complete) by a ceremony in April 1965. In the meantime, in November 1963 the Merritt Island facilities had been named the John F. Kennedy Space Center in memory of the assassinated president who inspired the nation to go to the Moon.

The design of the Launch Control Center built adjacent to the VAB was based on requirements specified by the Launch Operations Directorate, with the architectural work being performed by the firm of Urbahn–Roberts–Seelye–Moran of New York, and the construction being done by the same contractors as built the VAB. The four-storey structure measured 115 meters by 55 meters, and was 23.5 meters tall. It was designed not only for Apollo, but also with future manned programs in mind. There were four firing rooms, and launch crews faced the angled windows in the direction of

8-3 A Mobile Launch Platform with a Launch Umbilical Tower nears completion at the Kennedy Space Center in 1965. (NASA, Audin Malmin)

Launch Complex 39. These windows had large, top-to-bottom protective louvers that could close in a matter of seconds for protection against debris hurled out by an exploding rocket. Work on it began in March 1964, and was essentially complete by December 1965. The design of the building, both outside and in, was quite striking, and it won the 1965 Architectural Award for Industrial Design of the Year.

## Machines like none on Earth

At the core of the mobile launcher concept was a machine and a structure unlike anything built previously. The Mobile Launcher (ML) included the Mobile Launch Platform (MLP) and Launch Umbilical Tower (LUT). It would allow the assembly of the Saturn V inside the VAB, its transportation to the pad, propellant loading and monitoring of the stages and spacecraft. The design was done by Reynolds, Smith and Hills of Jacksonville, Florida. The Ingalls Iron Works Company of Birmingham, Alabama, constructed the platform and integrated tower, and the Brown Engineering

Company of Huntsville, Alabama, designed the enormous swing arms for the tower. Construction of the first of three mobile launchers at KSC began in July 1963. The first was topped out in September 1964, the second in December 1964 and the third in March 1965. The mechanical and electrical installation was done by Smith–Ernst of Orlando, Florida between December 1963 and March 1965. The swing arms were built and installed by the Hayes International Corporation of Birmingham, Alabama. By 1967, the ground support equipment had been supplied by the Pacific Crane and Rigging Company of Paramount, California.

The MLP measured 7.62 meters high, 41.15 meters wide and 48.77 meters long. It had a square hole, 13.7 meters per side, to pass the exhaust from the Saturn V's five F-1 engines to a deflector in a trench immediately below. It was an extremely rigid structure, with some steel sections more than 15 centimeters thick. Inside the VAB and on the pad, the platform rested on six pedestals. On the top of the platform at the exhaust opening were four massive hold-down arms that would support and secure the Saturn V. The LUT measured 121.5 meters from the deck of the platform to the top of the crane. There were nine swing arms, with the first to service the S-IC stage and the ninth to provide access to the spacecraft. The crane on top of the tower had a lift capacity of 22.7 metric tons at a distance of 15.24 meters from the swivel axis. The propellants were delivered to the S-IC stage from two different locations. RP-1 was stored in three 325,450 liter tanks approximately 400 meters from the pad. This was pumped to the pad, then through a connecting line in the MLP to the S-IC stage. LOX was drawn from a 21 meter diameter cryogenic tank and passed through a 35 centimeter line to the MLP, up the rear of the tower to the swing arm No. 1 level and then along the arm to the LOX inlet of the S-IC stage. Other arms serviced the upper stages.

Of all the ground-based machines and hardware of the Apollo program, one of the most astonishing is the Crawler-Transporter (C-T) to move either mobile launchers between the VAB and the pad, or the Mobile Service Structure that was added once a Saturn V had been delivered to the pad. After reviewing existing large machinery, NASA issued the contract to design and build the C-T to the Marion Power Shovel Company of Marion, Ohio. Fabrication and assembly of two C-Ts, each measuring 40 meters long by 34.5 meters wide, began in March 1963. The C-T had six massive hydraulic jacks for the pedestals of its payload. These could be continually adjusted to maintain the deck of the platform horizontal as it was driven up the shallow grade of the ramp to the pad. This system was so precise that it could maintain a Saturn V within ten minutes of arc of true vertical at all times! The C-T rode on four trucks, each having a diesel-powered generator to electrically drive its traction motor. Each truck had two belts of tread having 57 steel shoes, with each shoe weighing almost a metric tonne. After operational testing in Ohio, the C-Ts were disassembled, and the major components shipped to KSC and reassembled by the contractor. The first C-T moved under its own power there in January 1965, and in July it transported a ML. However, when metal fragments were discovered on the trucks, the bearings, rollers and related components had to be redesigned and modifications made to make them serviceable. Both C-Ts became operational in 1966. They have undergone scheduled maintenance and refurbishment and served

8-4 Two of the four massive hold-down arms on a Mobile Launch Platform which secured the S-IC stage of a Saturn V as the five F-1 engines built up to full thrust prior to lift off. (NASA, Audin Malmin)

America's space programs for more than four decades, being used throughout the Apollo, Skylab and Shuttle programs, and they will be used for the Constellation program.

## Launch Complex 39 special facilities and components

Often overlooked because it was rarely seen, was the ground support equipment that was an integral part of protecting Launch Complex 39 and of controlling the exhaust of the Saturn V during launch. The F-1 engines generated tremendous flame, smoke, pressure and sound, and all these had to be properly deflected. Beneath the hole in the MLP stood a large deflector in the flame trench of the concrete pad. The trench was 17.7 meters wide, 12.8 meters deep from the pad surface, and 137 meters long. Its walls and floor were lined with refractory brick capable of withstanding a flame of 3,000 degrees F moving at a speed of Mach 4. In the floor of the trench were rails to allow the travel and positioning of the flame deflector. This was a massive, heavy steel structure with two steeply sloping surfaces that were each curved at the bottom. It was marginally less than the width of the flame trench, stood 12.6 meters high and had a mass of 590 metric tons. After being positioned by a diesel locomotive, it was secured by thrust locks. There were two deflectors, and if one was damaged it could be replaced for the next launch.

To minimize the extent to which the exhaust from the F-1 engines could damage the ML, including the LUT swing arms, each pad had its own water supply. A pumping station located near the crawlerway to Pad 39B was capable of pumping water from holding tanks at a rate of 45,000 gallons of per minute to the deck of the mobile launcher, pad surfaces, flame trench walls, flame deflector, and swing arms of the tower. The 'pad deluge' was to soften the acoustic shock wave and prevent its reflection off the concrete from damaging the vehicle. Equipment on the deck of the platform also required protection from the F-1 engines. Three S-IC stage tail service masts provided electrical, hydraulic, propellant and pneumatic umbilical connections to the base of the S-IC stage up to the point of liftoff, at which time each mast would pivot up and away from the vehicle to an almost vertical position and a hood would cover its end to protect the exposed connectors. Also, once the four hold-down arms had released the Saturn V, blast hoods pivoted to protect them.

## APOLLO SATURN V LAUNCHES

In advance of initiating Saturn V flight operations, MSFC built and shipped to KSC the Facilities Checkout Vehicle, known as Saturn 500-F. The first stage, S-IC-F, had one F-1 mockup engine in an outboard mount and four dummy weights on the other positions. It left the Michoud Assembly Facility on the barge *Poseidon* on January 15, 1966, and arrived at the KSC turning basin near the VAB four days later. North American Aviation of Long Beach, California started to assemble the S-II-F in 1964, and on February 20, 1966 this set off by barge from Seal Beach on a journey down the Pacific coast, through the Panama Canal, aross the Gulf of Mexico, through the Florida Keys and up the Atlantic coast of Florida to KSC, arriving in the first week of March. The Douglas Aircraft Corporation manufactured the S-IVB-F in 1964. It left by barge on June 10, 1965 and arrived at KSC on June 30. The Apollo CSM for this checkout task was supplied by North American Aviation and flown in individual modules to KSC. The stacking of 500-F on ML No. 1 began on March 15, 1966, and it was transferred from the VAB to Pad 39A on May 25. However, when a hurricane threatened the area in June, 500-F was returned to the VAB for several days. Once it was back on the pad, the vehicle underwent months of facilities checks, including propellant loading and draining. After all required tests, the vehicle was returned to the VAB on October 14, 1966, and disassembled over the following week. The next Saturn V to be stacked on ML No. 1 would be flightworthy.

### Boeing and Rocketdyne activities in support of Saturn V launches

Boeing and Rocketdyne maintained field offices at KSC prior to, and in support of, Saturn V launches. Vince Wheelock recalled to this author the extent of activities by these companies at KSC during these years:

The integrated F-1 engine and S-IC stage launch processing tasks were done by the Boeing Company, with those involving F-1 engine tasks monitored and supported by Rocketdyne. F-1 engine-only launch processing tasks, including the F-1 engine ground support equipment operation, were conducted by a Rocketdyne team of 16 engineers, technicians, quality and supply personnel who reported to Lee Solid, the Resident KSC Rocketdyne Base Manager. For Saturn V launches, the Rocketdyne KSC launch processing team and NASA at KSC were technically supported on site by approximately five Rocketdyne program management and engineering specialists from Canoga Park. A key individual in this technical support was Bob Biggs, the manager of F-1 Flight Support.

During a typical Saturn V flight, there were approximately 25 operation and performance parameters measured on each F-1 engine, mostly included in the primary flight instrumentation system [comprised of] pressure transducers, temperature transducers, valve position indicators, flow measuring devices, vibration monitoring devices, power distribution junction boxes and associated electrical harnesses.

On Saturn V flights, KSC was responsible for the vehicle until it cleared the tower, and then MSC took over responsibility for the remainder of the flight [into orbit] and mission. Immediately after a launch, MSFC and Rocketdyne personnel at KSC conducted an analysis of the quick-look performance data. The F-1 engine primary data analysis was done at MSFC later, with NASA and Rocketdyne participating. It was done this way because MSFC had NASA design responsibility for the F-1 engine and, with Rocketdyne participating, was most technically qualified to analyze the F-1 engine flight performance. The primary detailed F-1 flight data review meetings were at MSFC, and there was participation in subsequent NASA Headquarters, MSC and KSC reviews as required.

Rocketdyne had a combined F-1 and J-2 engine shop in the VAB, on the west side. It housed the Rocketdyne technicians and inspectors, and had work benches, special test equipment, laminar flow equipment, tube brazers for the J-2 engine and small ground support equipment, but it could not accommodate an engine on its handler—if work was necessary on an engine on its handler, this was performed in the adjacent VAB high bay. Spare F-1 engines were not stored at KSC. If a replacement F-1 engine was required, it was brought from [...] the Michoud Assembly Facility. Only one F-1 engine (F-4023) on S-IC-3 was replaced at KSC.

## The launch of SA-501

The stages and other components for the first flight Saturn V began to arrive at KSC in August 1966. The S-IC was delivered on September 12, and erected on ML No. 1 in High Bay No. 1 on October 27. The stacking of SA-501 continued into 1967, and was followed by months of meticulous systems checks. It was moved to Pad 39A on August 26, the terminal countdown began on November 6, and the vehicle lifted off as scheduled on November 9, 1967. Although this was an unmanned launch, media representatives from around the world gathered at KSC to report the first mission of the enormous rocket designed to enable American astronauts to fly to the Moon. In

8-5 The S-IC stage of the 500-F Facilities Checkout Vehicle being outfitted in the transfer aisle of the Vehicle Assembly Building at the Kennedy Space Center. This Saturn V had one mockup F-1 outboard engine and four dummy weights in the other positions. (NASA/KSC)

8-6 A technician on a work platform inspects the mockup F-1 engine on the 500-F Facilities Test Vehicle. (NASA, Audin Malmin)

fact, the scale of the rocket was so staggering that thousands of Americans from all across the country arrived in campers, motor homes, cars, motorcycles and trucks to watch the launch. Every motel room from Daytona Beach down to Melbourne was booked. People from Rocketdyne, Boeing, Douglas, North American Aviation and many of their subcontractors were also there to witness this milestone in the Apollo program. The mission, designated Apollo 4, was an 'all up' test to prove not only the three stages of the launch vehicle, but also the Apollo spacecraft and the Worldwide Tracking Network. Two unplanned holds in the countdown amounted to almost four hours, but these were absorbed by 7.5 hours of built-in holds, and the launch was on schedule at 7:00 AM.

Essentially all of NASA's top brass were in the Launch Control Center, including Dr. Wernher von Braun. Dr. Kurt Debus, and Jim Murphy, who was Dr. Arthur Rudolph's Saturn V Deputy Program Manager. As Murphy recalled of that morning to film producer Mark Gray for the DVD collection, *The Mighty Saturns*:

> I went down for the launch and I was sitting next to Dr. von Braun. He said, 'Jim, why don't we run outside here before liftoff. I think you'd like it, and I know I'd like it.' So we ran outside and he said, 'You're going to really enjoy this.' And the first stage, of course, lit off and seven and a half million pounds of thrust with all the vibrations in the world were coming down on us. It set at your heart beat and

> . . . you've never felt anything like it before. You were in unison with the engines as they were going up. It was so perfect.

The ignition sequence start began with the center engine, then two of the opposing outboard engines and finally the remaining outboard engines, in what was known as a 1-2-2 engine start. At liftoff, the technicians in the Launch Control Center erupted in a chorus of cheers. As the F-1 engine nozzle extensions cleared the upper deck of the launch platform, the engines were gimbaled by a programmed yaw command to tilt the 36-storey vehicle away from the tower, lest one of the swing arms had been tardy in retracting. Once the vehicle was clear of the tower, it gimbaled the engines again to straighten up and continue its ascent. The cacophony of sound from the F-1 engines assaulted all those present. Rocketdyne and Boeing personnel were cheering SA-501 on, and more than a few wept as they witnessed six years of effort achieve fruition. The F-1 engines, with a combined output of 180 million horsepower, ran perfectly. The programmed Inboard Engine Cutoff occurred at 135.5 seconds. This was to reduce the loads on the launch vehicle, spacecraft and eventually astronauts prior to the remaining F-1 engines shutting down. Outboard engine cutoff occurred at 150.8 seconds, at which time the vehicle was traveling at 9,695 kph at an altitude of 63.7 km. Stage separation occurred at 151.4 seconds as the solid retro rockets at the base of the S-IC fired to retard the first stage and the ullage rockets of the S-II fired to move the second stage clear and settle its propellants prior to igniting its five J-2 engines. Its job complete, the S-IC fell back to Earth. Its telemetry was lost upon impacting the Atlantic Ocean at 410.0 seconds. The S-II and S-IVB stages suffered only minor anomalies, and the spacecraft performed its mission.

This was a tremendous start to the Saturn V flight program. Rocketdyne reported that the F-1 engines performed with only negligible departures from nominal thrust. On the ground, the pad inspection revealed expected and unexpected damage to the tower and platform. There had been fires at the hinge areas of swing arms 1, 2, 3 and 4, affecting components located there. The hoods for the tail service masts had been blown away by F-1 engine blast, and the exposed connectors, umbilical carriers and service lines damaged. The hoods protecting the hold-down arms were warped from the intense heat, and interior components damaged. Parts of the platform, including the F-1 engine service platform, were extensively damaged. Storage racks on three levels of the tower were badly damaged. Six of the hardened TV cameras positioned to record the launch up close had been totally destroyed, and four were damaged but repairable. Although the damage assessment led to upgraded protection and cooling of the platform and tower, it had always been recognized that some of the equipment on the pad would require repair or replacement after a launch.

### SA-502 sends a wakeup call

Attention then turned to SA-502, which was already on ML No. 2 in the VAB. The S-IC arrived at KSC on March 13, 1967 and was erected four days later. The S-II arrived in May. The completion of the stack was followed by months of methodical checking and testing. On February 6, 1968, it was rolled out to Pad 39A. Nearly two more months of tests and checks ensued. A NASA press release issued on March 28 scheduled the launch of Apollo 6 for April 3, but after the countdown rehearsal ran on to March 29 it was decided to slip the launch to April 4.

The objectives for this unmanned flight were to further validate the launch vehicle and its many systems, with additional checks of the spacecraft. In particular, the J-2 of the S-IVB stage was to restart in orbit. Of course, SA-502 did not attract the same degree of media interest as its predecessor. The countdown was completed without unplanned holds, and the vehicle lifted off at the set time of 7:00 AM on April 4. At approximately 110 seconds into the first stage burn, sensors detected a longitudinal oscillation in the launch vehicle known as 'pogo.' This continued until 140 seconds, then dissipated. It was later determined to have been generated by a combination of the launch vehicle structural longitudinal frequency and F-1 engine response to LOX suction lines resonant frequency. Inboard engine cutoff occurred at 144.7 seconds, outboard engine cutoff at 148.4 seconds, and stage separation shortly thereafter.

During the S-II burn, first one of the J-2 engines shut down, followed shortly after by a second engine. The remaining engines had to burn longer and, as a result of the manner in which the Saturn V's instrument unit was programmed, the vehicle ended up higher than planned at S-II shutdown, which caused the S-IVB to undergo radical maneuvers to steer towards the designed orbital insertion point. The S-IVB proved unable to restart its J-2 engine, and this prompted a revision of mission events and timing. Only nine of the 16 primary objects were fully achieved, six were partially achieved, and one (the J-2 restart) was not accomplished. All these problems had to be resolved before another Saturn V could be launched, and after investigations by MSFC and Rocketdyne they eventually were.

### SA-503: sending men to the Moon

The first manned launch of a Saturn V was Apollo 8, crewed by Commander Frank Borman, Command Module Pilot James Lovell, and Lunar Module Pilot William Anders. They had begun training in the summer of 1967 for an Earth-orbit mission, but in the first half of 1968 a bold plan began to take shape at NASA Headquarters in Washington and MSC in Houston. The mission was to have tested a lunar module in Earth orbit, but the development of this spacecraft was running seriously behind schedule. Why not launch Apollo 8 without a lunar module, and send the CSM on a mission to orbit the Moon. This would demonstrate that the S-IVB was able to make the translunar injection maneuver, test the ability of the CSM to function in cislunar space, and explore the navigational and tracking issues of a lunar mission. Another motivation for such an audacious flight was concern that the Soviets were preparing to send cosmonauts on a circumlunar loop, and it was considered important to beat the Soviets into cislunar space. When Borman's crew were informed on August 10, 1968 of the possibility a lunar orbit mission, they were enthusiastic and revised their training. The decision to proceed was made in October after Apollo 7 (launched by a Saturn IB) successfully demonstrated the CSM in Earth orbit.

8-7 An F-1 engine on a Vertical Engine Installer being delivered by flatbed truck to the Vehicle Assembly Building at the Kennedy Space Center. (NASA, Vince Wheelock Collection)

On arrival at KSC on December 27, 1967, the S-IC-3 stage was taken to the VAB. Engine number F-4023 was removed and replaced on May 31, 1968. Following the investigation into the 'pogo' suffered by Apollo 6, a system was installed in S-IC-3 to minimize this effect. This was described in the SA-503 Flight Evaluation Report, under A.4 Launch Vehicle as follows:

> A modification was made in the S-IC stage to suppress the longitudinal thrust oscillation observed in Apollo 6. A system was installed to provide gaseous helium to a cavity in each of the liquid oxygen pre-valves of the four outboard-engine suction lines. These gas-filled cavities act as a spring [...] to lower the natural frequency of the feed system and thereby prevent coupling between engine thrust oscillations and the first longitudinal mode of the vehicle structure.

SA-503 stacking began in mid-1968 with the erection of the S-IC-3 stage on the mobile launcher. The S-II arrived on June 26, and was stacked in July. The S-IVB arrived on September 12, and was added to the stack. Next was the Instrument Unit. Finally, the spacecraft was mated on October 7. Work continued around the clock to prepare the vehicle for the roll out to Pad 39A, which occurred two days later. After weeks of integrated systems testing, concluding with the Countdown Demonstration Test, the actual countdown was started, aiming for liftoff on 21 December.

With Apollo 8 being both the first manned Saturn V launch and the first manned mission to the Moon, media attention soared to a level not seen since the launch of John Glenn in 1962. The prime crew of Borman, Lovell and Anders, as well as the backup crew of Neil Armstrong, Buzz Aldrin and Fred Haise, continued to train up to the day before launch. Personnel from Rocketdyne and all the other contractors involved with Apollo were present for the launch. The countdown proceeded with no unplanned holds, and the Saturn V left the pad at 7:51:00 AM. In the Apollo 8 Mission Report written after the successful conclusion of the flight, Borman wrote in section 7.2.2. Powered Flight Phase, of the sensation of riding the Saturn V:

> Although the engine ignition sequence began 9 seconds before liftoff, no noise or vibration was apparent until 3 seconds before liftoff. From that time until liftoff, the primary sensations were a general vibration and an increasing noise level. Liftoff was easily discernible by the slight but sudden acceleration. From liftoff until the vehicle cleared the tower, the noise level continued to increase, and definite, but light, random lateral accelerations were sensed.
>
> The radio call for tower clear was received by the crew. However, the noise level continued to increase until any form of communication from the ground, or between the crew members, was impossible. This condition lasted for about 35 seconds, and could have prevented the crew from receiving an identifiable abort request call.
>
> The cabin pressure began to relieve as scheduled, accompanied by a rather loud surging noise. As the launch proceeded past 42 seconds, the noise level dropped off considerably and the remainder of the first stage operation was characterized by a smooth continuous increase in acceleration until inboard engine cutoff, when the acceleration leveled off at about 1 g. The first stage separation sequence was abrupt, with a sudden decrease in acceleration.

Once again, there were smiles at Rocketdyne and Boeing for the smooth operation of the S-IC stage. That was a portent for the mission. There were no anomalies in the second or third stage J-2 engine performance. When the crew began their orbit of the Moon and beamed TV images to Earth, it was a high point of the Apollo program. The crew read the moving opening passage from the *Book of Genesis*, then wished all those on Earth a Merry Christmas. After orbiting the Moon ten times in a 20 hour period, the Service Propulsion System of the Service Module was fired to break out of orbit and head home.

### SA-504 and Apollo 9

Apollo 9 was the first Saturn V to launch with a lunar module (LM). Its task was "to demonstrate the capability of the manned Apollo Command and Service Modules in earth orbit, particularly to evaluate LM systems capabilities, LM active rendezvous techniques, and combined CSM/LM functions." After the LM was retrieved from the S-IVB, this stage was to restart its J-2 engine as a further demonstration of its restart capability. The crew were Commander James McDivitt, Command Module Pilot David Scott and Lunar Module Pilot Russell Schweickart.

The first segment of SA-504 to arrive at KSC was the S-II stage, on May 15. The ascent and descent stages of the LM arrived June 14, the S-IVB on September 12, and S-IC-4 and IU both arrived on September 30. The vehicle was stacked on ML No. 2. The CSM arrived in October, and after checkout was added to the stack. The vehicle was rolled out to Pad 39A on January 3, 1969. Launch was initially intended for late February, but it was slipped several days owing to concerns about the crew's health. The countdown continued uneventfully, and SA-504 left the pad at precisely 11:00:00 AM on March 3. In the post-flight Mission Report, the Pilots' Report on the launch and powered flight was the essence subdued commentary:

> Some noise and vibration were apparent at ignition. The nature of this vibration changed slightly at liftoff, and some slight acceleration was felt. Immediately after liftoff, there was some vibration within the spacecraft, and the rate needles vibrated at a high frequency—up to about 1 deg/sec—in all three axes. The yaw, roll, and pitch programs all began and ended at the appropriate times.
>
> At approximately 50 seconds, the noise and vibration within the vehicle increased as the launch vehicle entered the region of maximum dynamic pressure. The noise and vibration levels remained reasonably high throughout this region, then decreased to very low levels for the remainder of powered flight. During the region of maximum dynamic pressure, the maximum indicated angle of attack was approximately 1 degree. Cabin pressure was relieved with an obvious, loud noise. Inboard engine cutoff occurred on time. Subsequently, a very slight chugging was felt for a short time, but damped out rapidly.
>
> Separation of the first stage proved to be quite a surprise. At engine shutdown, the acceleration felt in the couches changed very abruptly [...] and the Command Module Pilot and Lunar Module Pilot went forward against their restraint harnesses. The crew expected a decrease in acceleration to near zero, but not a change to negative.

The overall F-1 engine performance for the S-IC stage was 1.21 percent lower than predicted. However, at roughly 85 seconds engine No. 1 displayed an increase in thrust which was later determined to have resulted from fuel pump head decay. Although inboard engine cutoff occurred on time, outboard engine cutoff was 2.73 seconds later than predicted. Nevertheless, this did not affect the insertion into orbit of the spacecraft and the mission achieved virtually all of its objectives. The Boeing and Rocketdyne teams were satisfied with the performance of the S-IC stage, but Rocketdyne investigated the anomaly in the No. 1 engine.

## SA-505 and Apollo 10: rehearsal for the lunar landing

Flight hardware for the Apollo 10 mission began to arrive at KSC in October 1968. Interestingly, it was the LM that arrived first. It was sent to the altitude chamber of

8-8 The base of the 500-F Facilities Checkout Vehicle and one of the three tail service masts of a Mobile Launch Platform. (NASA, Audin Malmin)

the Manned Spacecraft Operations Building. The CSM arrived next, and it went into the altitude chamber as soon as the LM came out. The S-IC stage for SA-505 arrived on November 27, and soon thereafter was erected on MLP No. 3 in High Bay No. 2. The S-IVB arrived on December 3, and the S-II one week later. The Instrument Unit arrived on December 15. The spacecraft was transferred to the VAB on February 6, 1969 and added to the stack. On March 11, the vehicle was rolled out to Pad 39B, marking the first time that this had been used. The purpose of the mission was to test the LM in lunar orbit, as a full dress rehearsal to clear the way for the next mission to attempt a lunar landing. The crew was Commander Thomas Stafford, Command Module Pilot John Young and Lunar Module Pilot Eugene Cernan. The training was intensive, and starting in November 1968 they began a schedule of 12 hours per day spent in simulators, six days per week. The terminal countdown ran without holds to liftoff at 12:49 PM on 18 May.

The Pilots' Report from Apollo 10 Mission Report records the crew's impression of the Saturn V's first stage operation and other sights and sounds they experienced:

> Throughout the uneventful countdown, the Test Conductor [at KSC] and the crew maintained a timeline approximately 20 minutes ahead of the scheduled countdown activities. The final verbal count was initiated by the blockhouse communicator at 15 seconds prior to liftoff. Engine vibration and noise were first noted at 3.5 seconds before liftoff, then increased in magnitude until launch vehicle release, at which time the level decreased. The planned yaw maneuver started at 2 seconds with approximately two thirds of the magnitude experienced in simulators. Tower clearance was verified at approximately 12 seconds, followed by initiation of the programmed roll and pitch maneuvers. The roll program ended exactly at the predicted time. Noise and vibration levels again increased. However, these were less than had been experienced during a Gemini launch, and adequate intercommunications were maintained. Cabin pressure relieved at approximately 1 minute after liftoff. After [passing through] the maximum dynamic pressure region, the noise decreased to a low, steady roar. Inboard engine shutdown occurred on time and was accompanied by a slight longitudinal oscillation that damped rapidly. Outboard engine shutdown occurred at exactly 02:40:00 and was accompanied by longitudinal oscillations that damped after four cycles. The staging sequence and second-stage ignition occurred during these oscillations, and the appropriate engine lights were extinguished when the oscillations ended. The crew had expected a large negative pulse, and were therefore surprised by the series of rapid and relatively large fore-and-aft longitudinal oscillations.

F-1 performance for four of the five engines was within tenths or hundredths of one percent; that is to say, completely nominal. However, in the interval between 35 and 38 seconds, engine No. 1 displayed a drop in thrust to 6.611 MN as compared to the predicted 6.708 MN. At the time of outboard engine cutoff, its actual thrust was 6.761 MN. Nevertheless, the flight report stated that this did not significantly impair the flight.

Following translunar injection, the CSM separated and retrieved the LM from the S-IVB. Once in a circular lunar orbit at an altitude of 110 km, the LM separated and fired its descent engine to maneuver into an orbit with a low point 13 km above the surface on a track leading to the prime landing site. The descent stage was jettisoned and the ascent engine fired to enter an elliptical orbit in preparation for rendezvous with the CSM. The LM was released, and a programmed burn put it into a trajectory to orbit the Sun, so that it would not pose a collision hazard to future missions. Then the Service Propulsion System was fired to head home.

## SA-506 and Apollo 11's destiny at the Sea of Tranquility

In July 1969, three American astronauts became some of the best-known men of the 20th century. The public interest derived from the fact that their mission, Apollo 11, would be the first to attempt a manned lunar landing.

Almost half a million American men and women had worked for nearly a decade to meet President Kennedy's challenge. Apollo was not just about landing men on the

Moon and returning them safely to Earth. There were serious geopolitical stakes. America was proving to the world that it had the financial power and technical skill to achieve such a difficult goal.

All the elements of SA-506 arrived at KSC in January and February 1969, with LM-5 being first. The S-IC was erected on ML No. 1 on February 21. The CSM and LM underwent altitude chamber tests in March. Stacking was completed in April. The plugs-in test was conducted on May 14, and on May 20 the vehicle was rolled out to Pad 39A. The prime and backup crews continued to train up to the day before the launch, which was scheduled for July 16. Once again, reporters and other media from around the world congregated at KSC. There were no unplanned holds in the countdown, and at 9:32:00 AM the Saturn V with Commander Neil Armstrong, Command Module Pilot Michael Collins and Lunar Module Pilot Buzz Aldrin lifted off. The Pilots' Report of the launch was matter of fact:

> Liftoff occurred precisely on time, with ignition accompanied by a low rumbling noise and moderate vibration that increased significantly at the moment of hold-down release. The vibration magnitudes decreased appreciably at the time tower clearance was verified. The yaw, pitch and roll guidance program sequences occurred as expected. No unusual sounds or vibrations were noted while passing through the region of maximum dynamic pressure, and the angle of attack remained near zero. The S-IC/S-II stage sequence occurred smoothly and at the expected time.

Collins subsequently provided a visceral impression of the launch for the book *Apollo Expeditions to the Moon* (NASA SP-350):

> This beast is best felt. Shake, rattle, and roll! We are thrown left and right against our straps in spasmodic little jerks. It is steering like crazy, like a nervous lady driving a wide car down a narrow alley, and I just hope it knows where it's going, because for the first ten seconds we are perilously close to that umbilical tower.

From liftoff to outboard engine cutoff, the thrust of the F-1 engines averaged a mere 0.62 percent lower than predicted. Owing to additional measures to protect the mobile launcher, the damage was minimal. The S-II and S-IVB stages performed as expected.

After the LM *Eagle* landed safely on the Sea of Tranquility on July 20, 1969, the televised coverage of the moonwalk that followed became perhaps the most-watched broadcast in history. Tens of millions of people around the world watched the black and white image of Armstrong as he descended the ladder, stepped off the foot pad onto the surface and announced, "That's one small step for a man, one giant leap for mankind." It was one of America's proudest moments. Over the next 2 hours and 31 minutes, he and Aldrin set up instruments and collected 20 kg of samples for return to Earth. The total time spent by *Eagle* on the lunar surface was just over 21.5 hours. The ascent stage rendezvoused with Collins in the CSM *Columbia*. The ascent stage was discarded, to crash on the Moon. The spacecraft splashed down in the Pacific on

8-9 SA-501 nears completion of the stacking process inside the Vehicle Assembly Building at the Kennedy Space Center. (NASA/KSC)

July 24. After a period of quarantine against the possibility of lunar germs, the three men toured the world as good-will ambassadors.

With all the turmoil that had befallen the United States during the 1960s because of the Vietnam War and the murders of John F. Kennedy, Robert F. Kennedy and Martin Luther King, the success of the Apollo 11 mission was a shining moment of joy.

## SA-507 sends Apollo 12 to the Ocean of Storms

The goal of the Apollo 12 mission flown by Commander Charles "Pete" Conrad, Command Module Pilot Richard Gordon and Lunar Module Pilot Alan Bean was to land within walking distance of the Surveyor III spacecraft that landed in the Ocean of Storms in April 1967, and then make two moonwalks, the first to set up a suite of scientific instruments and the second to visit the unmanned craft.

The elements for SA-507 began to arrive at KSC in March 1969. The S-II arrived on April 21, the S-IC on May 3 and the S-IVB on May 9. The vehicle was stacked on ML No. 3, and rolled out to Pad 39A on September 8. It was overcast and raining at the Cape on the morning of the day of launch, November 14, but the countdown continued regardless. There were no unscheduled holds, and lift off occurred on time at 11:22 AM. Within seconds of clearing the tower, the vehicle disappeared into the low overcast. There was a heart-stopping moment in Mission Control at MSC when the Saturn V was struck by lightning. Fortunately, the F-1 engines continued to run without pause. The Pilots' Report recorded the events as they occurred:

> Engine ignition and liftoff were exactly as reported by previous crews. The noise level was such that no earpieces or tubes from the earphones were required. Communications, including the "tower clear" call, were excellent. A potential discharge through the space vehicle was experienced at 36 seconds after liftoff, and was noted by the Commander as an illumination of the gray sky through the rendezvous window, as well as an audible and physical sensing of slight transients in the launch vehicle. The master alarm came on immediately, and the following caution lights were illuminated: fuel cells 1, 2 and 3; fuel cell disconnect; main bus A and B undervoltage; ac bus 1; and ac bus 1 and 2 overloads. At approximately 50 seconds, the master alarm came on again, indicating an inertial subsystem warning light. Because the attitude reference display at the Commander's station was noted to be rotating, it was concluded that the platform had lost reference because of a low voltage condition. Although the space vehicle at this time had experienced a second potential discharge, the crew was not aware of its occurrence.
>
> The Lunar Module Pilot determined that power was present on both dc buses and had [...] 24 volts on both main dc buses. Although main bus voltages were low, the decision was made to complete the staging sequence before resetting the fuel cells to allow further troubleshooting by the crew and flight controllers on the ground. It was determined that no short existed, and the ground recommended that the fuel cells be reset. All electrical system warning lights were then reset when

> the fuel cells were placed back on line. The remainder of powered flight, through orbit insertion, was normal. The stabilization and control system maintained a correct backup inertial reference [and this] would have been adequate for any required abort mode.

Despite the two lightning strikes (or discharges of electrical potential as they were referred to) there was no significant damage to the launch vehicle or the spacecraft's instruments. The second discharge did affect the guidance system, but this was able to be reset later.

Vince Wheelock gave this author an explanation of the F-1 engines' continuing operation during and after the lightning strikes:

> Nominal F-1 engine performance occurred during the lightning strikes, and engine shutdown occurred as scheduled. F-1 engine data analysts reported no performance data change at the lightning strike times. The electrical system of the F-1 engine was rather simple, without computers that could be sensitive to high voltages. The electrical harnesses with the operational signals cross the interface from the S-IC stage to the F-1 engines. The engines were cocooned under thermal insulation, and the electrical harnesses were further protected by the supporting frames. The lightning strikes occurred after engine start, so the pyrotechnic igniters had already been fired. Shutdown occurs electrically by simultaneously energizing the engine control valve stop solenoid and the redundant shutdown valve solenoid. Had the lightning strike occurred while the Saturn V was on the launch pad, additional F-1 engine checkout would have been required per operating instructions in the technical manuals prior to launch.

After thorough checks of the spacecraft systems in Earth orbit, the crew was given permission to proceed with translunar injection. The intense training paid off when Conrad and Bean landed the LM *Intrepid* within 183 meters of Surveyor III. The first EVA lasted 4 hours, and saw the deployment of geophysical instruments and collection of soil and rock samples from nearby. They entered the LM and rested for about 7 hours and then ventured out again to conduct further experiments, perform a circuitous geological traverse to collect samples from a wider area and remove parts from Surveyor III. After just under 4 hours, they re-entered the LM and prepared for liftoff and rendezvous with *Yankee Clipper*. After the LM had been jettisoned, they headed home.

## SA-508 and the launch of Apollo 13

The S-IVB for SA-508 arrived at KSC on June 13, 1969, the S-IC on June 16, and the S-II on June 29. The spacecraft was not added to the stack until December, but finally, on December 15 the vehicle was rolled out to Pad 39A, where it faced four months of tests and checkouts.

The Apollo 13 crew was Commander James Lovell, Command Module Pilot John "Jack" Swigert and Lunar Module Pilot Fred Haise. In fact, Swigert was the

8-10 The crawler-transporter in the Vehicle Assembly Building bearing Mobile Launcher No. 1 and SA-501. (NASA/KSC)

backup CMP, and was brought in several days prior to launch to replace Ken Mattingly who had been exposed to German measles. The objective of the mission was to land near a crater in the Fra Mauro region of the Moon.

The countdown was uneventful, and SA-508 lifted off at 2:13:00 PM on April 11, 1970. The Pilots' Report in the Apollo 13 Mission Report states:

> Ignition and liftoff occurred on schedule. First stage performance was nominal and coincided very closely with simulations. Communications during the high noise level phase of flight were excellent. Staging of the S-IC occurred nearly on time and was accompanied by three distinct longitudinal oscillations.

The performance of all five F-1 engines in terms of stage thrust, specific impulse, total propellant consumption rate and total consumed mixture ratio were all within a fraction of one percent of predicted levels. In fact, total propellant consumption from hold-down arm release to outboard engine cutoff was low by a mere 0.06 percent.

However, one of the J-2 engines of the S-II shut down prematurely, requiring the remaining engines to burn longer. It was decided to proceed to the Moon. Almost 56 hours into the mission, with Apollo 13 in translunar coast, a cryogenic oxygen tank in the SM exploded and damaged the second tank positioned alongside. The story of the aborted mission has been well documented, and is one of the greatest stories of survival in the history of exploration, both on Earth and in space. Working with the flight control team, the astronauts displayed courage, ingenuity and calm to solve the many problems and splash down safely in the Pacific on April 17. Many years later, Lovell wrote a book of the mission that served as the basis of the movie *Apollo 13*, starring Tom Hanks. One point: although the ignition sequence of the F-1 engines was accurately portrayed, the film was inaccurate in a view looking down the length of the Saturn V because it depicted the swing arms of the LUT as moving one after the other, whereas in fact they moved simultaneously.

## SA-509 sends America's first astronaut to the Moon

On May 5, 1961, Alan Shepard became the first American astronaut to journey into space, flying a suborbital trajectory in the Mercury capsule *Freedom 7*. On acquiring an ailment of the inner ear, he was grounded in 1963 and appointed the Chief of the Astronaut Office. On later undergoing a surgical procedure to remedy his problem, he lobbied to command an Apollo mission, and was given Apollo 14. The Command Module Pilot was Stuart Roosa and the Lunar Module Pilot was Edgar Mitchell.

The S-IC, S-II and S-IVB stages for SA-509 all arrived at KSC on January 21, 1970. The Instrument Unit arrived on May 6. The vehicle was stacked on ML No. 2 in High Bay 3. On November 9, it was rolled out to Pad 39A. Launch was scheduled for January 31, 1971. After a 40 minute hold due to weather conditions in the final phase of the countdown, it was decided to proceed and to regain the time by revising the translunar trajectory. Lift off was at 4:23:02 PM. Interestingly, all that the Pilots' Report says of the S-IC stage's performance is:

> The proprioceptive cues reported by earlier crews were essentially unchanged during the launch of Apollo 14.

All F-1 engine performance parameters were well within one percent of the values predicted. Inboard engine cutoff was at 135.1 seconds, as planned. Outboard engine cutoff occurred at 164.1 seconds. This was the first flight of an S-IC stage to use a venturi in the LOX pressurization system instead of the GOX Flow Control Valve. It performed without problems.

The landing site was Fra Mauro, which had been missed by Apollo 13. Shepard and Mitchell set *Antares* down within 50 meters of the target point, and during just over 33 hours on the Moon they deployed a second suite of instruments and used the Modular Equipment Transporter to carry sampling tools and collect samples during a geological traverse that involved climbing a ridge. In all, they spent 9 hours and 25 minutes outside over two EVAs. At the end, Shepard produced a golf ball and tried a few strokes. On rendezvousing with *Kitty Hawk*, the ascent stage was jettisoned and the CSM headed home.

8-11 The turning basin at the Kennedy Space Center was the arrival point for S-IC and S-II stages arriving by barge. (NASA/KSC)

## SA-510 sends Apollo 15 and the Lunar Roving Vehicle to Hadley Rille

The mission of Apollo 15 was the first of the J-missions, with the new Lunar Roving Vehicle to expand the range of lunar exploration by the astronauts. The F-1 engines for the S-IC stage of SA-510 had been re-orificed to bring the tolerance closer to the target thrust. By this and other measures, such as stripping out some of the solid fuel retro rockets of the separation system, it had been possible to increase the propellant load of the LM for this mission to tackle a difficult landing site, and consumables to spend three days on the surface. The crew were Commander David Scott, Command Module Pilot Al Worden and Lunar Module Pilot James Irwin. In addition to flight training, Scott and Irwin trained using a 1-G LRV, and rehearsed sampling traverses using a field training vehicle nicknamed Grover. The LRV was engineered and built by Boeing in Washington State, with General Motors in Santa Barbara, California as the prime subcontractor. The manager at MSFC was Saverio Morea, who had played an important role in the development of the F-1 engine.

After the static hot fire test of the S-IC at the Mississippi Test Facility, F-1 engine No. 1 was replaced. However, based on the newly installed engine's own test history MSFC decided not to re-test the refitted stage. The S-II stage arrived at KSC on May 18, 1970, the S-IVB on June 13, the IU on June 26, and the S-IC on July 6. The S-IC was erected on ML No. 3 on 8 July, and the S-II and S-IVB stages and the IU were stacked in September. The LRV arrived in March 1971, and was installed on the LM in April. The spacecraft was added to the stack on May 8, and the completed vehicle was rolled out to Pad 39A on May 11. The launch was scheduled for July 26. The countdown had no unplanned holds. The typical method of starting the F-1 engines was to ignite the center engine first, followed by two pairs of opposing outboards in a sequence referred to as 1-2-2. However, owing to the changed F-1 engine on this S-IC stage, there was a low probability of this start sequence being achieved, and it was decided to use the sequence 1-1-2-1. Once all five F-1 engines had reached their full rated thrust, the vehicle lifted off at 8:34:00 AM as planned.

Lunar Module Pilot Jim Irwin wrote an autobiography, *To Rule The Night* (A. J. Holman Company, pub. 1973) about his years before, during and after his one and only space flight. He recorded his impressions of the launch:

> The time had been dragging, but the last minutes went very fast. Before we knew it, we heard the word "ignition." We sensed and then heard all that tremendous power being released underneath us on the pad. Slowly, tremulously, the rocket began to stir.
>
> We knew that if we cleared the tower we had a reasonable chance of survival if something should go wrong. I watched all the systems [that] I was responsible for on my side of the spacecraft. We cleared the tower. It was almost the happiest moment of my life to realize that after all those years it was now my turn. At last I was leaving the earth.

Scott had the advantage of having ridden a Saturn V on Apollo 9, as a point of comparison. The Pilots' Report recorded his impressions:

> Ignition and liftoff were positive with the same overall vehicle vibration frequency throughout S-IC flight that has been noted on previous flights. Noise levels were lower than those the Commander had experienced on Apollo 9, and communications were excellent throughout powered flight.

All F-1 engine performance parameters again remained within less than one half of one percent of those predicted. Inboard engine cutoff was at 136.0 seconds, and outboard engine cutoff at 159.5 seconds. The perfect performance of the F-1 engines was a metaphor for the rest of the mission. After landing on a confined plain in the mountainous Hadley–Apennine region of the Moon, Scott and Irwin deployed the LRV and traveled many kilometers from the LM *Falcon* during three EVAs. Their explorations were televised using the Lunar Communications Relay Unit by which the color TV camera on the LRV could be controlled from Earth.[1] After linking up with *Endeavour*, the ascent stage was discarded and the CSM headed for home.

### SA-511 and Apollo 16

It was ironic that the Apollo program was suffering budget cuts as the missions were beginning to deliver their greatest scientific return. After an investment of billions of dollars in infrastructure and in people, the Congress forced the cancellation of three missions. Nevertheless, the final two missions held the prospect of being the most exciting of the program. The crew of Apollo 16 certainly felt that way. John Young was Commander, Ken Mattingly was finally getting his shot at Command Module Pilot after having been pulled from Apollo 13. Charlie Duke was the Lunar Module Pilot.

The S-IC for SA-511 arrived at KSC on September 17, 1971, and was erected on ML No. 3 on September 21. The stacking operation was complete by December 8, and the vehicle was rolled out to Pad 39A five days later. The launch was scheduled for March 17, 1972. However, Apollo 16 had to be returned to the VAB to replace a critical component in the CSM, replace the parachute connector links (after one of the three parachutes on Apollo 15 had failed to deploy) and to resolve various other problems, as a result of which the launch date slipped to 16 April. The countdown proceeded smoothly, and SA-511 lifted off at 12:54:00 PM.

As the Pilots' Report in the Mission Report stated:

> The S-IC engine ignition and Saturn V liftoff were positively sensed. Vehicle vibration of the first stage was as reported on previous missions, and is probably best characterized as being similar to a freight train bouncing on a loose track. In-suit noise levels at maximum dynamic pressure were similar to the Apollo 10 levels recalled by the Commander. Communications were excellent throughout powered flight. Inboard engine cutoff on the S-IC stage was abrupt and was characterized by approximately four cycles of the S-II unloading. At S-IC outboard engine cutoff, the major four-cycle unloading of the S-II stage was again exhib-

[1] These stunning transmissions are now preserved on DVD by SpacecraftFilms.

8-12 S-IC-6 arrives at the Vehicle Assembly Building. It would become the first stage of SA-506 and be launched as Apollo 11 on July 16, 1969. (NASA/KSC)

> ited. S-IC outboard engine cutoff is the most impressive physiological experience of the Saturn V boost phase.

Duke wrote a much more visceral account of the launch in his book, *Moonwalker* (Oliver Nelson, pub. 1990):

> We heard a muffled roar from the huge engines below as they gulped down fuel at about 4,500 gallons per second. As the engines built up thrust for liftoff, I felt a tremendous vibration, and the spacecraft began to shake!
>
> I was startled. *Why is it shaking so hard?* I wondered anxiously. *What in the world is happening? There is something wrong with this thing.* I didn't recall any briefing to expect this—this violent vibration as we sat on the launch pad.
>
> John was saying, "We're go." The Cape was saying, "We're go." Everything is saying go, and I'm just sitting there shaking like crazy thinking. *This thing can't fly. It's going to shake to pieces.* If anyone wonders what an astronaut is doing at liftoff, this astronaut was *holding on*!

The performance of the F-1 engines was a carbon-copy of Apollo 15, being easily within the predicted parameters. As an example, total propellant consumption from hold-down arm release to outboard engine cutoff was low by an almost insignificant 0.51 percent. Overall, the stage's thrust was 0.05 percent higher than predicted. To use the vernacular, performance was 'nominal.'

In lunar orbit, the LM undocked as planned, but its landing was put on hold while a problem with the Service Propulsion System of *Casper* was resolved, then Young and Duke set *Orion* down on target in the Descartes–Caley area, and surprised their scientific mentors by finding the site to be very different to that expected.

## SA-512 takes Apollo 17 on the final manned voyage to the Moon

Apollo 17 was the final lunar landing mission of the program, and it drew almost as many visitors and members of the media to KSC as the first lunar landing mission, in part because it was to make a spectacular nighttime liftoff. The Commander was Eugene Cernan, the Command Module Pilot was Ronald Evans, and Dr. Harrison Schmitt, a professional geologist who trained as an astronaut, was the Lunar Module Pilot. Their target was a mountain valley in the Taurus–Littrow area.

The S-II for SA-512 arrived at KSC on October 27, 1970. The S-IVB followed on December 21. The spacecraft elements were delivered in 1971 and early 1972. The S-IC arrived in May 1972, and the Instrument Unit in June. The S-IC was erected on ML No. 3 on May 15, and the vehicle rolled out to Pad 39A on August 28. Some 23 hours prior to the scheduled time of launch, F-1 engine No. 2 lost its Gas Generator Igniter Installed indication. Rocketdyne dispatched technicians to replace both Gas Generator Igniters in that engine. The faulty igniters were later inspected, and it was determined that solder had been omitted from an electrical pin inside the igniter. The countdown continued aiming for liftoff late in the evening of December 6, until the Terminal Countdown Sequencer failed to issue the S-IVB LOX Tank Pressurization command at T-30 seconds, forcing a hold. Although the launch team worked around the problem, it imposed a delay of nearly three hours which slipped the launch time to 00:33:00 on December 7. All went well, however, and as the F-1 engines built up their thrust prior to liftoff, the blinding light turned night into day. As the vehicle climbed into the sky, it was visible to people hundreds of kilometers away.

The Pilots' Report stated:

> The crew could see first stage ignition through the rendezvous window and the hatch window cutout in the boost protective cover from a very few seconds prior to liftoff until a few seconds following liftoff. At staging of the S-IC stage, a bright flash was evident. It was as if the spacecraft was being overtaken by a fireball at the first stage cutoff.

The F-1 engines again proved their reliability by running within predicted levels. The stage thrust level was only 0.30 percent higher than that predicted, and the total propellant consumption was just 0.16 more than predicted. The IU initiated inboard engine cutoff at 139.3 seconds. The fuel depletion sensors initiated outboard engine

8-13 S-IC-13 is hoisted in the transfer isle of the Vehicle Assembly Building. It would form the first stage of SA-513 and launch Skylab. (NASA/KSC)

cutoff at 161.2 seconds. Rocketdyne and Boeing personnel at KSC gave a collective sigh of relief. The manned flight performance reliability of the F-1 engine had been 100 percent!

After *Challenger* landed on the Moon, the LRV was deployed and its TV camera sent back stunning color views. Dr. Schmitt described his findings with the various lunar rock and soil samples. After 75 hours on the Moon, Cernan and Schmitt lifted off and linked up with the CSM *America*. After the ascent stage had been jettisoned they headed home. Their splashdown in the Pacific Ocean on December 19 drew the Apollo lunar program to a close.

This was not the end of the F-1 engine's role, however. One more Saturn V was to be launched, this time carrying America's first space station.

## SA-513 launches the Skylab workshop

NASA began to investigate applications for Apollo-developed hardware in the early 1960s. Studies were conducted in the mid-1960s for an astronauts' workshop where scientific studies and experiments could be conducted. It was decided to base this on a spent S-IVB stage. It was a "wet workshop" because it would be launched as a live stage on a Saturn IB and outfitted in orbit by an Apollo crew. Later it was decided to outfit the S-IVB on the ground as a "dry workshop". However, since this would not be able to be launched as a live stage, it would have to be carried as the payload of a version of the Saturn V known as the Saturn INT-21. This two-stage launch vehicle would put the workshop into orbit. The astronauts would follow in an Apollo CSM launched by a Saturn IB.

The S-II for SA-513 arrived at KSC on January 1, 1971. The S-IC arrived on July 26, 1972, and was erected on ML No. 2 the following week. The S-II was stacked on September 20, 1972, the workshop was added on September 29, with the Instrument Unit following. An aerodynamic fairing was mounted on top. The plugs-in test was completed on March 21, 1973, and the vehicle rolled out to Pad 39A on April 26. The Countdown Demonstration Test was on May 2. There were no unplanned holds in the countdown, and SA-513 lifted off at 1:30:00 PM on May 14. Once again, the performance of the F-1s was nominal. The shutdown sequence had been changed from the 1-4 for manned flights to 1-2-2 for Skylab to reduce vehicle dynamics. The structural loads experienced during the S-IC boost phase were below design values. However, one of the workshop's solar array No. 1 covers suffered damage that later prevented it from hinging to deploy the array. The second solar array was ripped off completely, as was a large section of the thermal shield on the body of the workshop. Despite this setback, it proved possible to launch three crews to the workshop for durations, in turn, of 28, 59 and 84 days to undertake biomedical and scientific studies.

In all, 65 Rocketdyne F-1 engines powered 13 vehicles with a 100 percent success rate. This remains the high watermark for liquid propellant rocket engines that may never be seen again.

# 9

# The F-1A: the engine that might have been

The development of the F-1 engine occupied the decade from 1955 to 1965. As the work progressed, NASA and Rocketdyne realized that the lifting capability of the Saturn V for Apollo follow-on programs would be able to be increased by uprating the F-1 engine. The propellants would be RP-1 and either pure LOX or a blend of LOX and flourine, but since flourine was difficult to handle it would be used only as a last resort.

The proposal drawn up by Rocketdyne called for trying to produce 1.65 million pounds of thrust with minimum changes to the hardware, as an interim step toward an engine incorporating system optimization changes which would produce at least 1.8 million pounds of thrust.

## The 1650K F-1 engine

Rocketdyne started by extrapolating the mathematical model of the qualification engine redlines for the F-1 engine rated at 1.5 million pounds of thrust to achieve 1.65 million pounds, and then it redesigned such components as necessary to cope with the increased structural and performance redlines. This interim configuration was known as the 1650K engine. The engineers determined that this performance would be able to attained by replacing the 35-inch-diameter turbine with a 30-inch one spinning at a greater rate, and modifying the fuel and oxidizer pump inducers, thrust chamber structure, gas generator body and valve, and heat exchanger duct. This would boost the fuel manifold inlet pressure from the 1,534 to 1,697 psi, the LOX dome inlet pressure from a 1,438 to 1,596 psi, and the injector end pressure from 1,108 to 1,204 psi. The pump speed would increase from 5,291 to 5,645 rpm, the pump discharge pressure for LOX would rise from 1,556 to 1,735 psia, and the fuel discharge pressure would rise to 2,005 psia. The turbine flow would increase from 166 to 215 pounds per second, and the turbine would increase from 52,460 to 63,910 horsepower. No modifications would be made to the thrust chamber jacket, bands, tubes,

9-1 The F-1A engine held the promise of significantly increased sea level thrust and payload capacity. Its performance was proved by two development engines. (Rocketdyne, Harold C. Hall Collection)

the fuel manifold, LOX dome or the injector. The standard F-1 with a conservative sea level thrust of 1.52 million pounds provided 1.75 million pounds of thrust in the vacuum of space. The vacuum thrust of the 1650K engine would be 1.876 million pounds. A 13.50-inch model of the improved turbine was tested in May and June 1964, and its performance was greater than expected. The design of the 30-inch turbine was completed in 1965, and it was assembled and subjected to structural and hot fire tests in 1966. The uprated gas generator was also built and tested during this period.

Rocketdyne estimated that if it were to be authorized to proceed with the 1650K engine in mid-1965, it could deliver the first engine by the end of 1967 with flight qualification achieved by the end of 1968. Rocketdyne built at least one of these engines in order to test this rated configuration and provide the redlines necessary to proceed to the next level of engine performance.

## The F-1A engine

To increase the sea level thrust to the 1.8 million pound would require other design and structural changes. First, the pumps attached to the new 30-inch turbine had to be improved. The diameters of the LOX pump, LOX inducer and LOX propellant duct were all increased; as were the diameters of the fuel pump, impellers and fuel duct. The exhaust gas generator was redesigned to reduce the back pressure of the smaller turbopump, and a LOX pressure controller was added to the gas generator. The inlets, torus and flange of the LOX dome were all revised, as was the flange of the injector. The jacket, bands and control struts of the thrust chamber all had to be strengthened, as did the gimbal seat and the cross block. In the F-1A configuration, the turbopump spun at 5,625 rpm, and its pump discharge pressures were 1,887 psi for LOX and 2,240 psi for fuel. The turbine's flow was 225 pounds per second and it produced an impressive 75,720 horsepower. The uprated gas generator now had 1,247 psi at the injector, a LOX valve inlet pressure of 1,540 psi, and a fuel valve inlet pressure of 1,870 psi. The specific impulse of 269.6 was within five seconds of the original engine. The F-1A weighed only about 1,500 pounds more than the production model of the orginal engine. Rocketdyne estimated that if it were to be given the go-ahead in June 1965, the first engine could be delivered by the end of 1968 and flight qualified by the end of 1969. The company set out to produce two F-1A engines (designated F-10404 and F-109-4) in order to prove their performance.

The F-1A went through the same development program as the original engine, but at a more rapid pace owing to the manufacturing and testing experience of the F-1 and the fact that some components were off the shelf. In tests at Edwards Field Laboratory the new engines performed as expected. However, Rocketdyne did not receive the hoped-for contract to manufacture the F-1A for future missions. After rising steeply in the early 1960s, NASA's funding for research and development of Saturn launch vehicles peaked in 1966 at $1.63 billion and then fell dramatically. By 1969 it was doubtful that all of the Saturn V launch vehicles ordered would be required, and there were no programs in prospect that would produce payloads so heavy as to require such a powerful vehicle.

9-2 Detail of the F-1A engine.

## Rocketdyne's Knowledge Retention Program

The F-1 rocket engine was by far the most ambitious of the programs undertaken by Rocketdyne, and preserving the knowledge base was essential. In 1969, therefore, it set out to formalize the research, design, development and production experiences of the managers. A volume was compiled for each of the major components, detailing the tooling, planning documents, non-standard machines, methods and procedures for fabrication, cleaning and processing procedures, production personnel, purchase labor, purchase parts, the criteria for make or purchase, problems encountered and solutions achieved, producibility actions, and the steps that would be necessary to restart the program.

## The F-1A engine and the Space Exploration Initiative

In 1969, as Apollo was achieving John F. Kennedy's goal, it was decided to develop a fully reusable Space Shuttle. But financial constraints imposed compromises, and when the design was announced in 1971 it was for a partially reusable vehicle. In an effort to make this cost effective, it was decided to phase out all expendable rockets and rely upon the Shuttle as the National Space Transportation System. (When this decision later proved to be an error, it was reversed.) The long-term objective of the Shuttle was to ferry into orbit the elements required to assemble a large station. But this posed the question of whether America intended to restrict its future operations to low orbit, or to follow up Apollo by establishing a base on the Moon and sending a manned mission to Mars.

Studies were made in the 1980s into how America might resume its exploration of space, and on July 20, 1989, on the 20th anniversary of Apollo 11's lunar landing, President George H. W. Bush stood on the steps outside the National Air and Space Museum in Washington D.C. and called for a Space Exploration Initiative (SEI). This would be:

> ... a long-range continuing commitment. First, for the coming decade, for the 1990s, Space Station Freedom, our critical next step in all our space endeavors. And, for the next century, back to the Moon, back to the future, and this time, back to stay. And then a journey into tomorrow, a journey to another planet, a manned mission to Mars. Each mission should and will lay the groundwork for the next.

The National Space Council, chaired by Vice President Dan Quayle, formulated the ambitious plan to accomplish this in terms of time, technology and funds. NASA Administrator Richard Truly directed Aaron Cohen, Director of the Johnson Space Center in Houston, Texas, to undertake a 90-day study to define the major elements of the SEI. Truly presented the *Report of the 90-Day Study on Human Exploration of the Moon and Mars* to the National Space Council in November. Section 5 of the report described the necessary launch vehicles:

> [The existing] Space Shuttle and expendable launch vehicle fleet will support robotic missions; Space Shuttle will be used for crew transport to Space Station Freedom.
>
> For establishing the lunar outpost, an Earth-to-orbit lift capability of approximately 60 metric tons is required. The only vehicle in this class being considered before 1999 is the Shuttle-C; after that time, use of the new Advanced Launch System vehicles is possible.
>
> For establishing the Mars outpost, an Earth-to-orbit lift capability of approximately 140 metric tons is required. A larger Shuttle-derived heavy-lift vehicle and an Advanced Launch System vehicle are being considered for this purpose.
>
> For either Shuttle-derived or Advanced Launch System heavy-lift vehicles, some degree of enhancement of ground launch and production facilities will be necessary.

The Shuttle-C (for cargo) used three Space Shuttle Main Engines (SSMEs) and uprated solid rocket boosters. The Advanced Launch System (ALS) would have a 10 meter diameter core stage powered by five SSMEs, and would be augmented by four advanced solid rocket boosters. There were alternative configurations; one being an 'all liquid' monster in which the core stage with three SSMEs would be augmented by three boosters each of which would have six SSMEs, for a total of 21 SSMEs!

SEI was very ambitious and very expensive, with the price tag ranging upwards of $500 billion over a period of 20 years. For it to get the support of Congress and the American people, it would require to be scaled back. In an effort to obtain a broader consensus, NASA established Project Outreach. It gathered proposals from industry, private institutions, advocacy groups and individuals, and in 1990 NASA formed the Synthesis Group to study these submissions and make recommendations for SEI.[1] In May 1991 the Synthesis Group released its report, *America at the Threshold*. This found the earlier figures for payloads to low Earth orbit to have been conservative, and said the payload capability for a lunar mission should be 150 metric tons and the payload for a flight to Mars should be 250 metric tonnes. The report also contained some surprising recommendations regarding a Heavy Lift Launch Vehicle (HLLV). Under the *Options* section of *Supporting Technologies*, it stated:

> Current heavy lift launch vehicle concepts are concentrating on liquid oxygen–liquid hydrogen for main booster engines, together with advanced solid rocket motors for liftoff augmentation. Experience has shown that large liquid hydrogen and oxygen engines have been expensive to develop and operate. A liquid oxygen–hydrogen propellant is not an attractive option for the first stage of a heavy lift launch vehicle because of the large tank volume and the safety concerns of using hydrogen below an altitude of 100,000 feet. A new launch vehicle capability will need to be developed for both the Space Exploration Initiative and other Department of Defense and NASA [...] requirements; alternate propulsion system concepts should be considered for these applications.

[1] The Synthesis Group was also known as the Stafford Committee because it was headed by former Apollo astronaut Tom Stafford.

> The Apollo Saturn V launch vehicle program developed the F-1 liquid oxygen–kerosene (RP-1) powered booster engine [...] which delivered 1.5 million pounds of thrust at sea level. Flight reliability was demonstrated to be 100%. Although it was never flown, an upgraded version delivered 1.8 million pounds of thrust at sea level. The potential exists for a Heavy Lift Launch Vehicle booster to support the Space Exploration Initiative using proven and reliable technology. The use of F-1 engines as a first stage and strap-on propulsion stage of a new Heavy Lift Launch Vehicle is extremely attractive from cost, schedule and safety viewpoints. Using F-1s for booster engines, coupled with liquid oxygen-hydrogen upper stage engines (upgraded J-2s or space transportation system main engines), could result in establishing Heavy Lift Launch Vehicle capabilities by 1998. The Soviet Union, which currently has heavy lift capabilities of approximately 100 metric tonnes to low Earth orbit, has relied on liquid oxygen-kerosene technology (RD-170 engine) for booster stages with liquid oxygen-hydrogen for the upper stages.

Rocketdyne was one of the corporations which contributed to Project Outreach. It had recommended the F-1 engine as a flight-proven booster for SEI. The Synthesis Group concurred, and unequivocally stated that the F-1A engine would be the best engine to power the large boosters envisioned for SEI. It was not just the F-1's flight history which prompted the Synthesis Group to endorse the F-1A. Rocketdyne had twice demonstrated its ability to restart an engine program.

## The F-1A production plan

In 1991, after the Synthesis Group sought information to assess Rocketdyne's ability to restart the F-1 program, the company reviewed its Knowledge Retention Program. The part and assembly drawings and tooling drawings would require revision to current standards. Much of the original tooling had been scrapped, but new tooling would exploit advances in manufacturing and assembly, and also cut labor costs. It was estimated that $100 million would be required to retool for the F-1A program, with a total non-recurring cost of $500 million. Some 60 percent of Rocketdyne's suppliers for the F-1 program were still in business, and others had been established to support Rocketdyne's Space Shuttle Main Engine program. Most important, more than 300 Rocketdyne personnel involved in the design, testing and manufacturing of the F-1 were still employed by the company or could be recalled from retirement to act as consultants. Rocketdyne also had the advantage of residual F-1 hardware. Five spare flight engines were in bonded storage at the Michoud Assembly Facility near New Orleans, Louisiana. They had all been test fired, and could be used as design and manufacturing references and even as test articles, if necessary, once portions were disassembled, inspected, and necessary parts replaced.[2] Other F-1 engines were on

[2] These F-1 engines were: F-4023 with six tests for a total of 456.1 seconds; F-5036, three, 249.6 seconds; F-6045, three, 309 seconds; F-6049, eight, 501.1 seconds; and F-6090, three, 250 seconds.

9-3 A Rocketdyne technician together with the F-1A 30-inch turbopump (left) and F-1 35-inch turbopump. (Rocketdyne, Harold C. Hall Collection)

display around the United States, and could be commandeered if necessary. The two F-1A engines that Rocketdyne had built and test fired were also available, and would be the primary production reference.

The F-1A of the 1990s would benefit from advances in manufacturing technology such as Computer Aided Design (CAD), Computer Aided Machining (CAM), CNC (Computer Numerically Controlled) machining, automation, robotics, laser cutting, multi-axis machining, inspection of parts, alloys and other materials, and improved process control and documentation resulting from the advent of personal computers and more powerful mainframe computers. The thrust chamber tube forming, furnace braze and vacuum welding capability created for the F-1 had been maintained and upgraded to support other engine programs. In the years since F-1 production ended, tube forming, stacking and furnace brazing had improved to such a degree that the second brazing cycle had been eliminated. Certain machined parts were now made with nearnet castings, and in many cases one-piece castings could replace multipart assemblies, thereby offering considerable manufacturing cost savings.

Most of the dedicated F-1 engine test stands at the Edwards Field Laboratory had been deactivated, but could be reactivated at a fraction of the cost of their original construction. The Air Force said Test Stand 1-D or 1-E could be made operational for $20 million. Test Stand B-2, used for the Space Shuttle Main Propulsion Test Article at NASA's Stennis Space Center, could be converted for F-1A testing. Once the dedicated F-1 engine test stand at MSFC had been refurbished, it could be used to test the uprated F-1.

## SEI launch vehicles powered by the F-1A

In March 1992 a paper entitled *The Saturn V F-1 Engine Revisited* was given at the Space Programs and Technologies Conference of the American Institute of Aeronautics and Astronautics. The authors were Billy Shelton, the Program Manager of Expendable Launch Vehicle Applications at MSFC, and Terry Murphy, Chief of the Systems Engineering Division for Program Development at Rocketdyne. The paper described the F-1 engine performance and history, the F-1A configuration for the 1990s, and the requirements to restart production. It showed how the F-1A would be able to support SEI, in particular describing a vehicle with a core stage of the same diameter as a Shuttle propellant tank that was powered by four SSMEs and had two strap-on boosters each of which had a pair of F-1As. Such a configuration would be able to put 270,000 pounds on a translunar trajectory. A version with the same core and four boosters would be able to send 400,000 pounds to Mars. Two other sample configurations were a core stage with four SSMEs and four boosters which had one F-1A, and the same core stage with eight such boosters. Four months later, Murphy gave a paper entitled *The F-1A Engine: Cost Effective Choice for Heavy Lift Launch Vehicles* at a Joint Propulsion Conference in Nashville, Tennessee, reinforcing the case for the F-1A as the logical choice to power the large boosters needed for SEI. This time he included a Saturn V-derived vehicle with a payload shroud of the same 33-foot diameter as the S-IC and S-II stages. The core booster would be powered by five F-1A engines. The two boosters would each have two F-1A engines. The liftoff weight of the vehicle was estimated at 13.3 million pounds, and the thrust at launch at 16.2 million pounds. Such vehicles could be assembled in the VAB, although new mobile launchers and launch umbilical towers would be required, and after various modifications it would be possible to launch them from Launch Complex 39.

## First lunar outpost and the F-1A-powered HLLV

In the summer of 1992 NASA unveiled a rather less ambitious program to return to the Moon. This reflected a push within NASA to try to develop a more cost-effective means of resuming human lunar exploration than that of the SEI. The launch vehicle which NASA illustrated leaving the VAB for this new scheme, dubbed First Lunar Outpost, was essentially that described in the AIAA paper by Shelton and Murphy, but with the core stage powered by five F-1A engines, and each of the two boosters having a pair of F-1A engines. This vehicle was configured to provide a new means of sending a payload to the Moon. Instead of lunar orbit rendezvous, this approach was 'Lunar Direct'. In some respects, it harkened back to the dual-launch concept of using two Saturn Vs for Apollo. The HLLV vehicle, sometime referred to as Comet, would launch an unmanned habitat module to a specified site on the Moon, where it would make an automated landing. It would deploy solar panels, activate the habitat and await a crew. Another Comet would launch a four-man crew with an integrated lander and return stage. After setting down alongside the habitat, they would power down their craft and transfer to the habitat. About 45 days later, they would re-enter their craft, and ascend into lunar orbit as a preliminary to heading home. However,

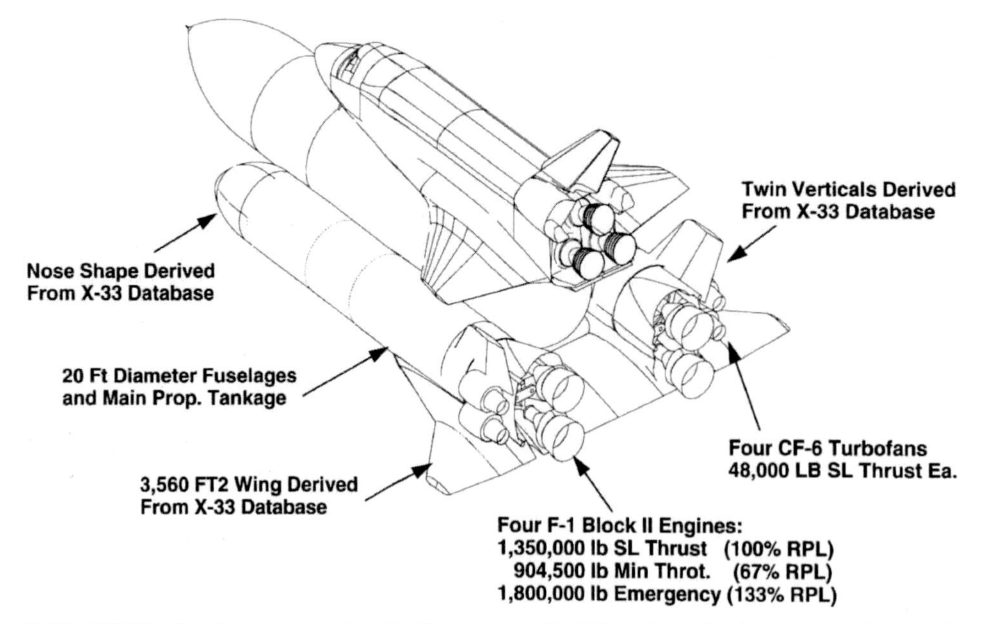


9-4 In 1996 Rocketdyne proposed to NASA using the F-1 Block II engine in the Liquid Fly Back Booster as a replacement for the Shuttle's solid rocket boosters. With recovery and refurbishment, the liquid booster would have been capable of being used twenty times. (Rocketdyne, Vince Wheelock Collection)

this scheme followed SEI into oblivion. The 1992 presidential election was won by Bill Clinton, who had expressed no interest in human space exploration. At that time NASA's budget was devoted primarily to the flying the Shuttle and designing Space Station Freedom. For the foreseeable future, therefore, there would be no funding for lunar exploration. It appeared that the last hope for resurrecting the F-1 in the form of the F-1A had gone.

## Liquid Rocket Booster and Liquid Flyback Booster studies

A number of conceptual studies in the 1990s focused on liquid rocket boosters using F-1A engines in support of the Shuttle. *Liquid Rocket Booster Feasibility Study for the Space Shuttle* was published by L.M. Rivera of Texas A & M University in 1994 as AIAA-95-0007. This considered a common liquid rocket booster employing the F-1A engine as a replacement for the solid rocket boosters of the Shuttle and also for the HLLV. This would use a single F-1A engine, and be both throttleable and able to be shut down. In 19961997, Rocketdyne investigated using the F-1A (which had been renamed the F-1 Block II) as a propulsion system of a Liquid Flyback Booster for the Shuttle, in the event that NASA decided to pursue such an option to replace the solid rocket boosters. The engine would combine aspects of the original F-1 and the F-1A, and be throttleable—a capability that was demonstrated on F-1A engine

F-109-4. When NASA decided to continue to use the solid rock booster, Rocketdyne terminated this study.

## THE END OF AN ERA

The Rocketdyne F-1 engine was an amazing engineering achievement of the 1960s, and remains one of the mechanical marvels of the 20th century. Its flight reliability of 100 percent was achieved through sound engineering, exhaustive testing, constant improvement, and the personal sacrifices of the many people who worked on it. The F-1 engine dispatched the Apollo lunar missions which marked some of the proudest moments in American history. There are F-1 engines on display, both outdoors and indoors, at museums across the United States. Perhaps the finest is at the Space and Rocket Center in Huntsville, Alabama. This pristine engine is complete, right down to the red protective covers and safety wired hex nuts. It rests on its Air Transport Engine Handler, seemingly as if it had just had its final inspection at Rocketdyne. In April 2007, I had the privilege of standing alongside this engine with Sonny Morea, who managed the F-1 program at MSFC. He had not seen this particular display before. He pointed out the various parts of the engine and then, after a moment of silent reflection, acknowledged the contributions of those at Rocketdyne and the Marshall Space Flight Center who had made it a reality, with the simple phrase:

"What a magnificent engine!"

# Appendix

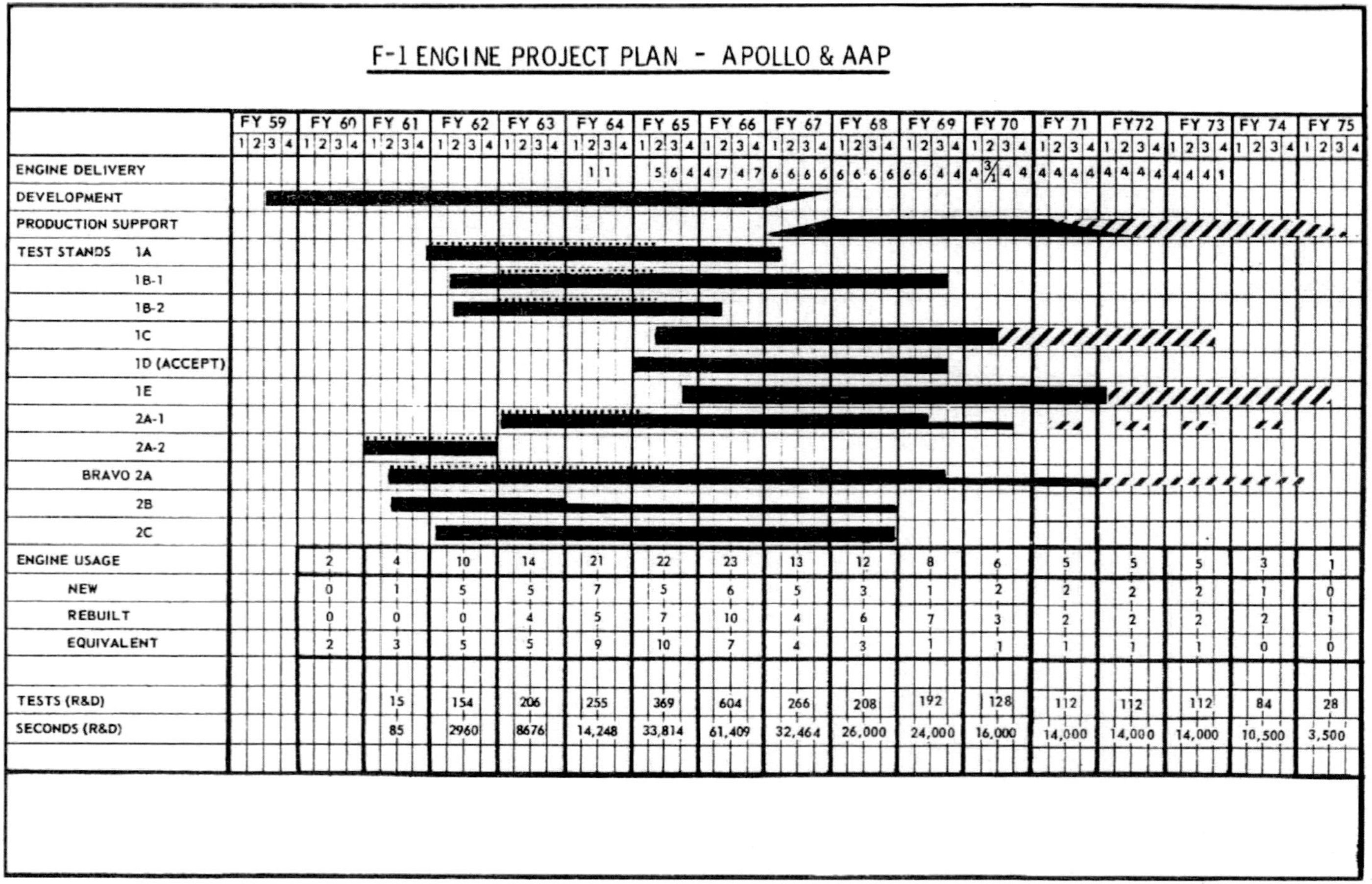

## F-1 ENGINE PROJECT PLAN - APOLLO & AAP

| | FY 59 | FY 60 | FY 61 | FY 62 | FY 63 | FY 64 | FY 65 | FY 66 | FY 67 | FY 68 | FY 69 | FY 70 | FY 71 | FY72 | FY 73 | FY 74 | FY 75 |
|---|---|---|---|---|---|---|---|---|---|---|---|---|---|---|---|---|---|
| | 1 2 3 4 | 1 2 3 4 | 1 2 3 4 | 1 2 3 4 | 1 2 3 4 | 1 2 3 4 | 1 2 3 4 | 1 2 3 4 | 1 2 3 4 | 1 2 3 4 | 1 2 3 4 | 1 2 3 4 | 1 2 3 4 | 1 2 3 4 | 1 2 3 4 | 1 2 3 4 | 1 2 3 4 |
| ENGINE DELIVERY | | | | | | 1 1 | 5 6 4 | 4 7 4 7 | 6 6 6 6 | 6 6 6 6 | 6 6 4 4 | 4 3/1 4 4 | 4 4 4 4 | 4 4 4 4 | 4 4 4 1 | | |
| DEVELOPMENT | | | | | | | | | | | | | | | | | |
| PRODUCTION SUPPORT | | | | | | | | | | | | | | | | | |
| TEST STANDS 1A | | | | | | | | | | | | | | | | | |
| 1B-1 | | | | | | | | | | | | | | | | | |
| 1B-2 | | | | | | | | | | | | | | | | | |
| 1C | | | | | | | | | | | | | | | | | |
| 1D (ACCEPT) | | | | | | | | | | | | | | | | | |
| 1E | | | | | | | | | | | | | | | | | |
| 2A-1 | | | | | | | | | | | | | | | | | |
| 2A-2 | | | | | | | | | | | | | | | | | |
| BRAVO 2A | | | | | | | | | | | | | | | | | |
| 2B | | | | | | | | | | | | | | | | | |
| 2C | | | | | | | | | | | | | | | | | |
| ENGINE USAGE | | 2 | 4 | 10 | 14 | 21 | 22 | 23 | 13 | 12 | 8 | 6 | 5 | 5 | 5 | 3 | 1 |
| NEW | | 0 | 1 | 5 | 5 | 7 | 5 | 6 | 5 | 3 | 1 | 2 | 2 | 2 | 2 | 1 | 0 |
| REBUILT | | 0 | 0 | 0 | 4 | 5 | 7 | 10 | 4 | 6 | 7 | 3 | 2 | 2 | 2 | 2 | 1 |
| EQUIVALENT | | 2 | 3 | 5 | 5 | 9 | 10 | 7 | 4 | 3 | 1 | 1 | 1 | 1 | 1 | 0 | 0 |
| TESTS (R&D) | | | 15 | 154 | 206 | 255 | 369 | 604 | 266 | 208 | 192 | 128 | 112 | 112 | 112 | 84 | 28 |
| SECONDS (R&D) | | | 85 | 2960 | 8676 | 14,248 | 33,814 | 61,409 | 32,464 | 26,000 | 24,000 | 16,000 | 14,000 | 14,000 | 14,000 | 10,500 | 3,500 |

I-E-F 2306 B SEPTEMBER 22, 1967

## F-1 ENGINE/J-2 ENGINE COST HISTORY

| | FY 59 | FY 60 | FY 61 | FY 62 | FY 63 | FY 64 | FY 65 | FY 66 | FY 67 | FY 68 | FY 69 | FY 70 | | TOTAL |
|---|---|---|---|---|---|---|---|---|---|---|---|---|---|---|
| TOTAL COST | 3,578 | 22,152 | 45,859 | 81,922 | 116,935 | 195,894 | 266,235 | 286,698 | 232,734 | 175,916 | 117,201 | 63,573 | | 1,608,697 |
| ROCKETDYNE | 3,578 | 22,152 | 44,404 | 77,149 | 104,580 | 175,579 | 237,006 | 242,872 | 200,894 | 150,932 | 96,850 | 50,630 | | 1,406,626 |
| PROPELLANT | | | 1,455 | 4,773 | 12,355 | 20,315 | 29,229 | 39,492 | 26,111 | 17,369 | 13,972 | 7,509 | | 172,580 |
| AEDC | | | | | | | | 4,334 | 5,729 | 7,615 | 6,379 | 5,434 | | 29,491 |
| F-1 ENGINE | 3,578 | 22,152 | 37,121 | 50,958 | 59,202 | 100,614 | 137,424 | 129,876 | 101,720 | 78,830 | 45,125 | 18,604 | | 785,204 |
| R'DYNE | 3,578 | 22,152 | 36,081 | 48,075 | 55,638 | 94,709 | 125,329 | 112,339 | 92,990 | 72,613 | 40,856 | 16,717 | | 721,077 |
| DEV./PROD. SUP/O&FS | 3,578 | 22,152 | 36,081 | 48,075 | 47,267 | 60,580 | 61,181 | 48,112 | 29,124 | 28,150 | 20,719 | 9,492 | | 414,511 |
| $NAS_W$-16 | 3,578 | 22,152 | 36,081 | 48,075 | 47,267 | 60,580 | 61,181 | 48,112 | 10,427 | 798 | 175 | 123 | | 338,549 |
| NAS8-18734A | | | | | | | | | 18,697 | 27,352 | 20,544 | 9,369 | | 75,962 |
| PRODUCTION 98 | | | | | 8,371 | 34,129 | 64,148 | 64,227 | 63,866 | 44,463 | 20,137 | 7,225 | | 306,566 |
| NAS8-5604 76 | | | | | 8,371 | 34,129 | 64,148 | 64,227 | 53,168 | 13,092 | 1,413 | 1,518 | | 240,066 |
| NAS8-18734B 22 | | | | | | | | | 10,698 | 31,013 | 17,935 | 5,676 | | 65,342 |
| NAS8-18734F | | | | | | | | | | 358 | 789 | 11 | | 1,158 |
| PROPELLANT | | | 1,040 | 2,883 | 3,564 | 5,905 | 12,095 | 17,537 | 8,730 | 6,217 | 14,269 | 1,887 | | 64,127 |
| DEV./PSP/O&FS | | | 1,040 | 2,883 | 3,564 | 5,450 | 9,680 | 15,193 | 7,005 | 5,286 | 3,583 | 1,706 | | 55,390 |
| PRODUCTION | | | | | | 455 | 2,415 | 2,344 | 1,725 | 931 | 686 | 181 | | 8,737 |
| J-2 ENGINE | | | 8,738 | 30,964 | 57,733 | 95,280 | 128,811 | 156,822 | 131,014 | 97,086 | 72,076 | 44,969 | | 323,493 |
| AEDC | | | | | | | | 4,334 | 5,729 | 7,615 | 6,379 | 5,434 | | 29,491 |
| R'DYNE | | | 8,323 | 29,074 | 48,942 | 80,870 | 111,677 | 130,533 | 107,904 | 78,319 | 55,994 | 33,913 | | 685,549 |
| DEV./PROD. SUP./O&FS | | | 8,323 | 29,074 | 37,762 | 46,370 | 47,538 | 53,905 | 41,256 | 39,124 | 31,016 | 22,110 | | 358,478 |
| NAS8-19A | | | 8,323 | 29,074 | 37,762 | 46,370 | 47,538 | 53,905 | 41,256 | 39,124 | 12,566 | 531 | | 318,449 |
| NAS8-19C | | | | | | | | | | | 18,450 | 21,579 | | 40,029 |
| PRODUCTION 152 | | | | | 9,180 | 34,500 | 64,139 | 76,628 | 66,648 | 39,195 | 24,978 | 11,803 | | 327,071 |
| NAS8-5603 103 | | | | | 9,180 | 34,500 | 64,139 | 74,733 | 23,571 | 3,[illegible] | 266 | 275 | | 210,050 |
| NAS8-19B 49 | | | | | | | | 1,895 | 43,077 | 35,[illegible]09 | 24,712 | 10,761 | | 116,254 |
| NAS8-19K | | | | | | | | | | | | 767 | | 767 |
| PROPELLANT | | | 415 | 1,890 | 8,791 | 14,410 | 17,134 | 21,955 | 17,381 | 11,152 | 9,703 | 5,622 | | 108,453 |
| DEV./PSP/O&FS | | | 415 | 1,890 | 8,791 | 13,643 | 13,601 | 15,481 | 11,779 | 8,583 | 7,448 | 4,684 | | 86,315 |
| PRODUCTION | | | | | | 767 | 3,533 | 6,474 | 5,602 | 2,569 | 2,255 | 938 | | 22,138 |

# F-1 ENGINE PRODUCTION COST HISTORY

| | FY 63 | FY 64 | FY 65 | FY 66 | FY 67 | FY 68 | FY 69 | FY 70 | | TOTAL | UNIT |
|---|---|---|---|---|---|---|---|---|---|---|---|
| ENGINES | | 2 | 15 | 23 | 23 | 19 | 12 | 4 | | 98 | 1 |
| TOTAL COST | 8,371 | 34,129 | 64,148 | 64,227 | 63,866 | 44,463 | 20,137 | 7,225 | | 306,566 | 3,128 |
| FEE (7.977%) | | 2,687 | 4,023 | 4,554 | 5,467 | 3,817 | 1,925 | 174 | | 22,647 | 231 |
| DELIVERABLE HARDWARE | 8,141 | 29,504 | 57,232 | 58,104 | 55,787 | 37,598 | 14,931 | 3,497 | | 264,794 | |
| ENGINES | 6,562 | 24,734 | 48,726 | 51,144 | 52,308 | 33,540 | 13,455 | 3,134 | | 233,603 | 2,384 |
| 1ST 14 ENGINES | 6,562 | 16,670 | 5,157 | 554 | 159 | 32 | 166 | 270 | | 29,570 | |
| 1ST 76 ENGINES | | | | | | | | 207 | | 207 | |
| ENGINE SYSTEMS | | 1,795 | 6,961 | 8,555 | 8,076 | 6,270 | 2,246 | 601 | | 34,504 | |
| ENG. SUPPORT SERS. | | | | | 535 | 1,472 | 1,198 | 315 | | 3,520 | |
| MAINT. ENGR. | | | | | | 1,054 | 1,299 | 977 | | 3,330 | |
| RUT | | | | | 2 | 43 | 33 | 7 | | 85 | |
| QAT | | | | | | | 39 | 29 | | 68 | |
| THRUST CHAMBERS | | 1,634 | 9,028 | 10,926 | 13,181 | 5,660 | 1,985 | 173 | | 42,582 | |
| GAS GENERATORS | | 134 | 1,130 | 1,298 | 1,519 | 897 | 452 | 32 | | 5,462 | |
| TURBOPUMPS | | 1,097 | 5,861 | 8,937 | 9,999 | 4,622 | 2,171 | 179 | | 32,866 | |
| PROP. FEED SYSTEMS | | 744 | 3,837 | 4,981 | 4,985 | 2,780 | 1,079 | 108 | | 18,514 | |
| HYD. CONTROL SYS. | | | | | 237 | 931 | 457 | 21 | | 1,696 | |
| PRESSURIZATION SYS. | | 99 | 848 | 2,014 | 2,338 | 1,849 | 320 | 46 | | 7,514 | |
| ELEC. & FLIGHT INSTR. | | 274 | 1,416 | 2,577 | 2,234 | 3,206 | 779 | 62 | | 10,548 | |
| THERMAL INSULATION | | | 2,502 | 5,951 | 5,889 | 2,739 | 545 | 32 | | 17,658 | 180 |
| TOOLING | | 1,672 | 9,263 | 3,861 | 2,644 | 1,683 | 647 | 64 | | 19,844 | 202 |
| STE | | 615 | 2,728 | 1,490 | 460 | 292 | 39 | 11 | | 5,635 | 58 |
| PARTS | 25 | 2,951 | 5,831 | 4,971 | 2,921 | 3,331 | 617 | 340 | | 20,987 | 214 |
| MOCKUPS | 1,400 | 343 | 426 | 16 | 8 | (9) | | | | 2,184 | 22 |
| GSE | 154 | 1,476 | 2,249 | 1,973 | 550 | 401 | 122 | 13 | | 6,938 | 71 |
| LONG LEAD HARDWARE | | | | | | 335 | 737 | 10 | | 1,082 | |
| SITE ASSOC. SUPPORT | 230 | 1,938 | 2,893 | 1,569 | 2,612 | 3,048 | 3,281 | 3,554 | | 19,125 | 195 |
| FIELD ENGR | 97 | 150 | 356 | 722 | 1,028 | 1,447 | 1,768 | 1,658 | | 7,226 | |
| SUPPORT PARTS SERS. | 50 | 146 | 368 | 497 | 1,115 | 859 | 889 | 1,464 | | 5,388 | |
| TRAINING | 33 | 60 | 59 | 20 | 22 | 23 | 24 | 25 | | 266 | |
| MANUALS | 50 | 274 | 269 | 237 | 265 | 338 | 313 | 311 | | 2,057 | |
| OTHER | | 1,308 | 1,841 | 93 | 182 | 381 | 287 | 96 | | 4,188 | |
| L.S. ACTIVIATION | | 1,308 | 1,841 | | | | | | | 3,149 | |
| DATA EVALUATION | | | | 32 | 51 | 115 | | | | 198 | |
| NEW TECHNOLOGY | | | | 61 | 131 | 242 | 238 | 46 | | 718 | |
| SPECIAL EFFORT | | | | | | 24 | 49 | 50 | | 123 | |

# Index

**Plate 1.** An F-1 thrust chamber on Test Stand 2-A prior to test at the Edwards Field Laboratory. (Rocketdyne, Frank Stewart Collection)

**Plate 2.** An F-1 engine on a flatbed truck about to be installed on a test stand at the Edwards Field Laboratory in 1962. (Rocketdyne, Vince Wheelock Collection)

**Plate 3.** The Engine Rollover Adapter Dolly and Rollover Sling were used to position the F-1 engine for installation on a test stand. (Rocketdyne, Vince Wheelock Collection)

**Plate 4.** The Lift Table beneath the Engine Rollover Adapter Dolly was used to raise the F-1 into its final position to be secured on the test stand. (Rocketdyne, Vince Wheelock Collection)

**Plate 5.** A mainstage test of an F-1 on Test Stand 1-A at the Edwards Field Laboratory in 1961. (Rocketdyne, Vince Wheelock Collection)

**Plate 6.** For a few seconds in its ignition sequence, the F-1 was engulfed in flames until the hot gas flow of mainstage carried this inferno away. (Rocketdyne, Vince Wheelock Collection)

**Plate 7.** An F-1 in full mainstage on one of the development test stands at the Edwards Field Laboratory in 1962. (MSFC)

**Plate 8.** The retort containing an F-1 thrust chamber continues to glow from the intense heat after the brazing cycle on December 12, 1961. (Rocketdyne, Harold C. Hall Collection)

**Plate 9.** F-1 engine No. 2091 being assembled in December 1968. (Rocketdyne, Harold C. Hall Collection)

**Plate 10.** A finished F-1 engine at Rocketdyne in January 1968. (Rocketdyne, Harold C. Hall Collection)

**Plate 11.** The opposite side of the F-1 in the previous illustration. (Rocketdyne, Harold C. Hall Collection)

**Plate 12.** An F-1 thrust chamber undergoing a Panoramic Gamma Ray Weld Test to verify the integrity of the weldments after the second furnace brazing. (Rocketdyne, Harold C. Hall Collection)

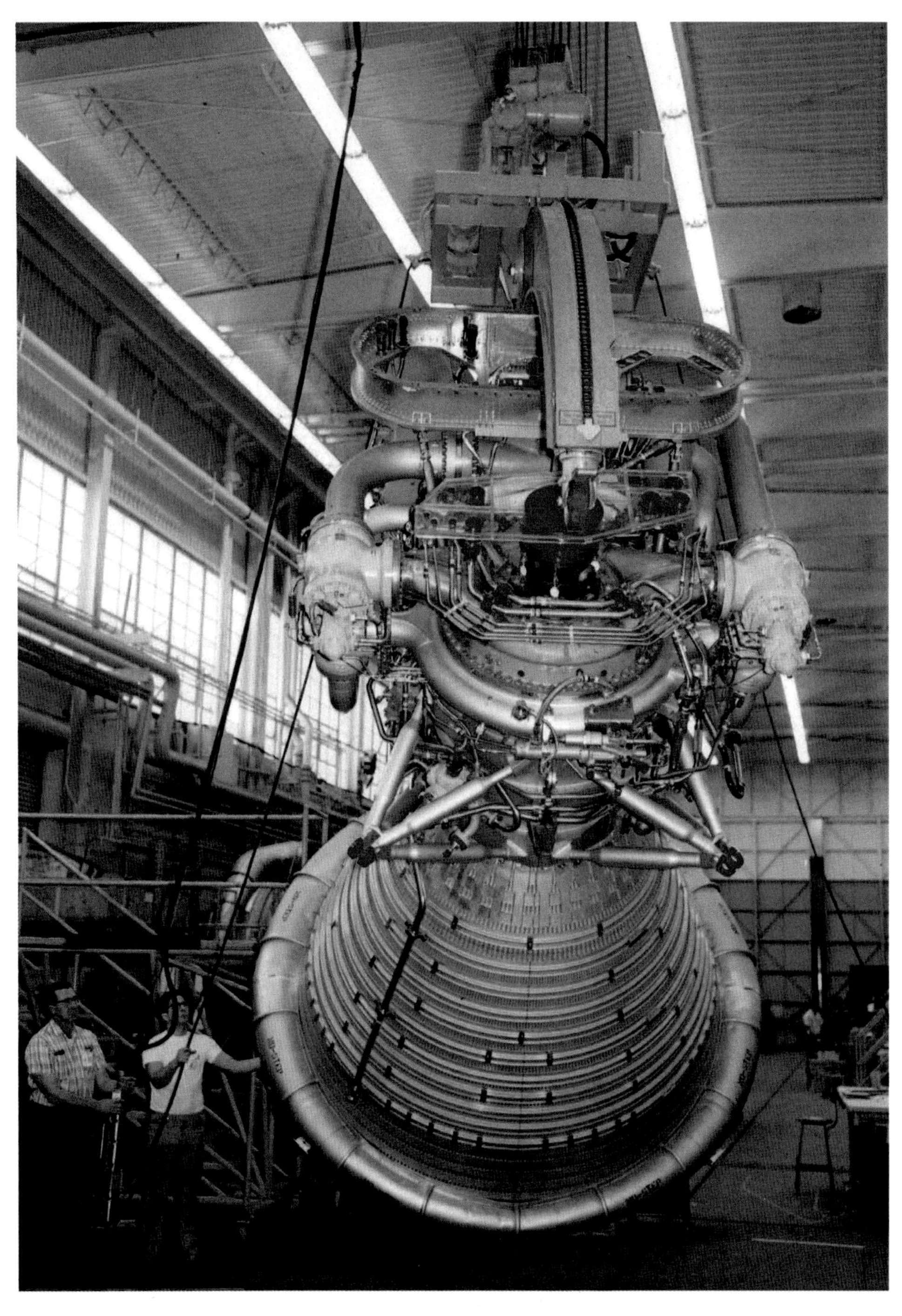

**Plate 13.** An Engine Rotating Sling is being used to move and reorient a finished F-1 engine in the Rocketdyne plant. (Rocketdyne, Vince Wheelock Collection)

**Plate 14.** Finished F-1 engines and F-1 thrust chambers shipped to the Marshall Space Flight Center were stored in the F-1 Engine Preparation Shop of Building 4666, shown here in March 1965. Note the configuration of the propellant lines. (NASA/MSFC)

**Plate 15.** The center F-1 engine is installed on the S-IC-T stage in the S-IC Static Test Stand at the Marshall Space Flight Center in March 1965. (NASA/ MSFC)

**Plate 16.** Installing a nozzle extension on an F-1 engine on S-IC-T. (NASA/MSFC)

**Plate 17.** A completed F-1 engine installed on the S-IC-T stage in the S-IC Static Test Stand at the Marshall Space Flight Center. The first hot fire test of all five engines on S-IC-T was on April 16, 1965. (NASA/MSFC)

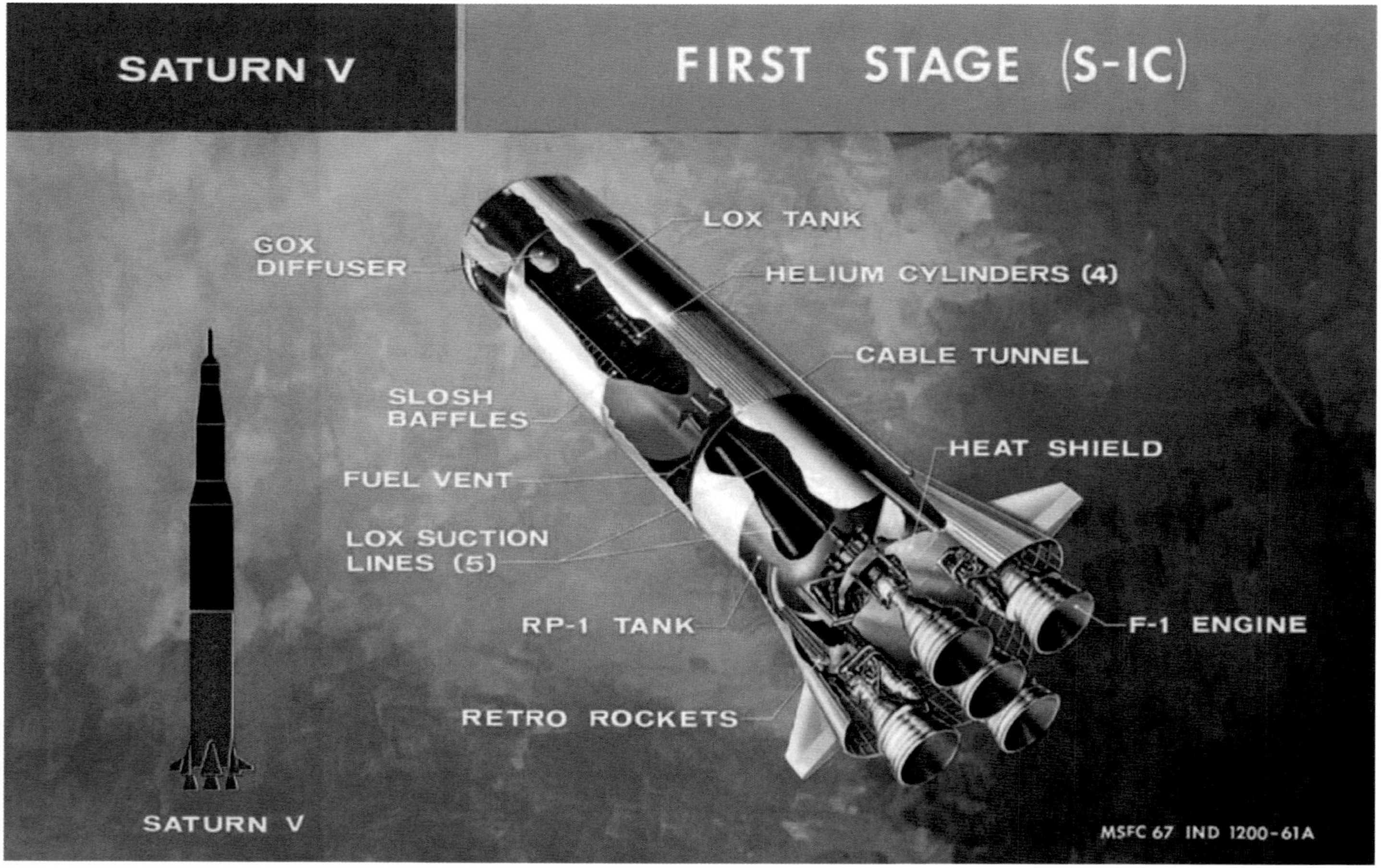

**Plate 18.** A cutaway illustration of the S-IC stage issued in 1967. (NASA/MSFC)

**Plate 19.** The second S-IC stage built and tested at the Marshall Space Flight Center is rolled out from the post-static test refurbishment facility in 1966. (NASA, Vince Wheelock Collection)

**Plate 20.** A crane lowers the thrust structure of S-IC-9 into the assembly stand in the Vertical Assembly Building at the Michoud Assembly Facility in October 1967. (NASA/MSFC)

**Plate 21.** A Boeing technician prepares an F-1 engine for mounting on the thrust structure of S-IC-8 in October 1967. (NASA/MSFC)

**Plate 22.** A specialized crane was used to mount F-1 engines on the S-IC stage thrust structure at the Michoud Assembly Facility. (NASA/MSFC)

**Plate 23.** Four S-IC stages in various stages of assembly inside the massive Michoud Assembly Facility in the fall of 1968. (NASA/MSFC)

**Plate 24.** A dawn view of SA-501 on Pad 39A. When launched on November 9, 1967 for the Apollo 4 mission it was the first flight test of the cluster of F-1 engines. (NASA/MSFC)

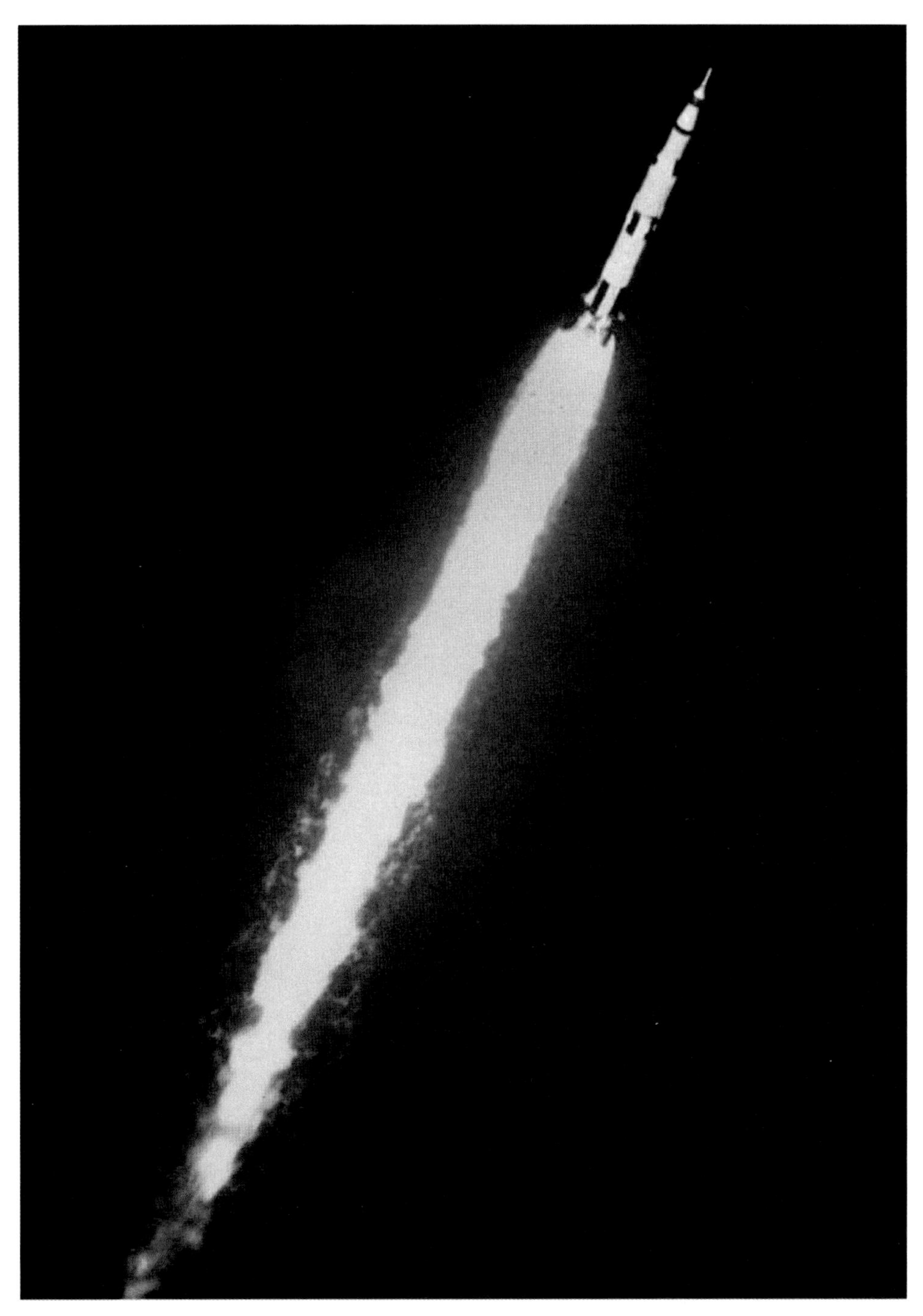

**Plate 25.** A ground tracking camera followed SA-501 during its flawless ascent. The F-1 engines provided a flaming exhaust trail that was more than four times the length of the enormous vehicle. (NASA/ MSFC)

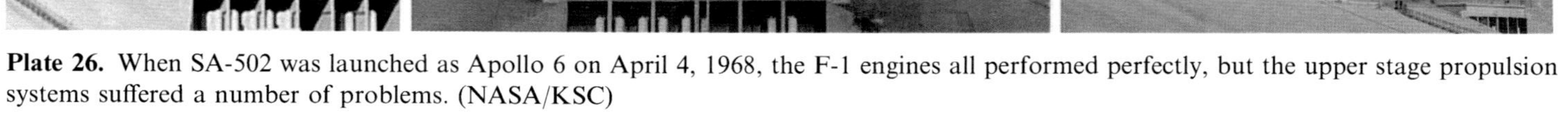

**Plate 26.** When SA-502 was launched as Apollo 6 on April 4, 1968, the F-1 engines all performed perfectly, but the upper stage propulsion systems suffered a number of problems. (NASA/KSC)

**Plate 27.** S-IC-3 in the transfer aisle of the Vehicle Assembly Building. It formed the first stage of SA-503 and became the first manned Saturn V when it was launched as Apollo 8. (NASA/KSC)

**Plate 28.** A million people were drawn to the Kennedy Space Center to watch the launch of Apollo 11 on July 16, 1969. (NASA/KSC)

**Plate 29.** The last Saturn V to be launched was SA-513, in a one-off configuration with America's first space station, Skylab in the third stage position. (NASA/KSC)

**Plate 30.** An F-1A engine photographed in December 1969. (Rocketdyne, Harold C. Hall Collection)

**Plate 31.** The exhaust gas flow straightener of the 30-inch-diameter turbine of the F-1A engine. (Rocketdyne, Harold C. Hall Collection)

**Plate 32.** The F-1A turbopump featured a smaller turbine and numerous other changes and refinements designed to increase the engine's sea level thrust. (Rocketdyne, Harold C. Hall Collection)

Printing: Mercedes-Druck, Berlin
Binding: Stein+Lehmann, Berlin

Made in the USA
Las Vegas, NV
04 September 2021

29573323R00200